Praise for the first edition

"This book is a wonderful tool with which to think.
It offers an expansive introduction to the field of science
studies, a rich exploration of the theoretical terrains it
comprises and a sheaf of well-reasoned opinions that
will surely inspire argument."
Geoffrey C. Bowker, University of California, San Diego

"Sismondo's *Introduction to Science and Technology Studies*, . . .
for anyone of whatever age and background starting
out in STS, must be the first-choice primer: a resourceful,
enriching book that will speak to many of the successes,
challenges, and as-yet-untackled problems of science studies.
If the introductory STS course you teach does not fit
his book, change your course."
Jane Gregory, *ISIS*, 2007

An Introduction to Science and Technology Studies

Second Edition

Sergio Sismondo

This second edition first published 2010
© 2010 Sergio Sismondo

Edition history: Blackwell Publishing Ltd (1e, 2004)

Blackwell Publishing was acquired by John Wiley & Sons in February 2007. Blackwell's publishing program has been merged with Wiley's global Scientific, Technical, and Medical business to form Wiley-Blackwell.

Registered Office
John Wiley & Sons Ltd, The Atrium, Southern Gate, Chichester, West Sussex, PO19 8SQ, United Kingdom

Editorial Offices
350 Main Street, Malden, MA 02148-5020, USA
9600 Garsington Road, Oxford, OX4 2DQ, UK
The Atrium, Southern Gate, Chichester, West Sussex, PO19 8SQ, UK

For details of our global editorial offices, for customer services, and for information about how to apply for permission to reuse the copyright material in this book please see our website at www.wiley.com/wiley-blackwell.

The right of Sergio Sismondo to be identified as the author of this work has been asserted in accordance with the Copyright, Designs and Patents Act 1988.

Library of Congress Cataloging-in-Publication Data

Sismondo, Sergio.
　An introduction to science and technology studies / Sergio Sismondo. – 2nd ed.
　　p.　cm.
　Includes bibliographical references and index.
　ISBN 978-1-4051-8765-7 (pbk. : alk. paper)
　1. Science–Philosophy.　2. Science–Social aspects.　3. Technology–Philosophy.
4. Technology–Social aspects.　I. Title.
　Q175.S5734 2010
　501–dc22　　　　　　　　　　　　　　　　　　　　　　　　　　　2009012001

A catalogue record for this book is available from the British Library.

Set in 10/12.5pt Galliard by Graphicraft Limited, Hong Kong

11　2017

Contents

Preface

Science & Technology Studies (STS) is a dynamic interdisciplinary field, rapidly becoming established in North America and Europe. The field is a result of the intersection of work by sociologists, historians, philosophers, anthropologists, and others studying the processes and outcomes of science, including medical science, and technology. Because it is interdisciplinary, the field is extraordinarily diverse and innovative in its approaches. Because it examines science and technology, its findings and debates have repercussions for almost every understanding of the modern world.

This book surveys a group of terrains central to the field, terrains that a beginner in STS should know something about before moving on. For the most part, these are subjects that have been particularly productive in theoretical terms, even while other subjects may be of more immediate practical interest. The emphases of the book could have been different, but they could not have been very different while still being an introduction to central topics in STS.

An Introduction to Science and Technology Studies should provide an overview of the field for any interested reader not too familiar with STS's basic findings and ideas. The book might be used as the basis for an upper-year undergraduate, or perhaps graduate-level, course in STS. But it might also be used as part of a trajectory of more focused courses on, say, the social study of medicine, STS and the environment, reproductive technologies, science and the military, or science and public policy. Because anybody putting together such courses would know how those topics should be addressed – or certainly know better than does the author of this book – these topics are not addressed here.

However the book is used, it should almost certainly be alongside a number of case studies, and probably alongside a few of the many articles mentioned in the book. The empirical examples here are not intended to replace rich detailed cases, but only to draw out a few salient features. Case studies are the bread and butter of STS. Almost all insights in the field grow out of them, and researchers and students still turn to articles based on cases to learn central ideas and to puzzle through problems. The empirical examples used in this book point to a number of canonical and useful studies. There are many more among the references to other studies published in English, and a great many more in English and in other languages that are not mentioned.

This second edition makes a number of changes. The largest is reflected in a tiny adjustment of abbreviation. In the first edition, the field's name was abbreviated S&TS. The ampersand was supposed to emphasize the field's name as Science and Technology Studies, rather than Science, Technology, and Society, the latter of which was generally known as STS in the 1970s and 1980s. When the ampersand seemed important, the two STSs differed considerably in their approaches and subject matters: Science and Technology Studies was a philosophically radical project of understanding science and technology as discursive, social, and material activities; Science, Technology, and Society was a project of understanding social issues linked to developments in science and technology, and how those developments could be harnessed to democratic and egalitarian ideals. When the first edition of this book was written, the ampersand seemed valuable to identifying its terrain. However, the fields of STS (with or without ampersand) have expanded so rapidly that the two STSs have blended together. The first STS (with ampersand) became increasingly concerned with issues about the legitimate places of expertise, about science in public spheres, about the place of public interests in scientific decision-making. The other STS (without) became increasingly concerned with understanding the dynamics of science, technology, and medicine. Thus, many of the most exciting works have joined what would once have been seen as separate. This edition, then, increases attention to work being done on the politics of science and technology, especially where STS treats those politics in more theoretical and general terms. As a result, the public understanding of science, democracy in science and technology, and political economies of knowledge each get their own chapters in this edition, expanding the scope of the book.

Besides this large change, there is considerable updating of material from the first edition, and there are some reorganizations. In particular, the chapter on feminist epistemologies of science has been brought forward, to put it in better contact with the chapters on social constructivism and the strong programme. The four chapters on laboratories, controversies, objectivity, and creating order have been reorganized into three.

I hope that these additions and changes make the book more useful to students and teachers of STS than was the first. It is to all teachers and students in the field, and especially my own, that I dedicate this book.

Sergio Sismondo

1

The Prehistory of Science and Technology Studies

A View of Science

Let us start with a common picture of science. It is a picture that coincides more or less with where studies of science stood some 50 years ago, that still dominates popular understandings of science, and even serves as something like a mythic framework for scientists themselves. It is not perfectly uniform, but instead includes a number of distinct elements and some healthy debates. It can, however, serve as an excellent foil for the discussions that follow. At the margins of science, and discussed in the next section, is technology, typically seen as simply the application of science.

In this picture, science is a formal activity that creates and accumulates knowledge by directly confronting the natural world. That is, science makes progress because of its systematic method, and because that method allows the natural world to play a role in the evaluation of theories. While the scientific method may be somewhat flexible and broad, and therefore may not level all differences, it appears to have a certain consistency: different scientists should perform an experiment similarly; scientists should be able to agree on important questions and considerations; and most importantly, different scientists considering the same evidence should accept and reject the same hypotheses. The result is that scientists can agree on truths about the natural world.

Within this snapshot, exactly how science is a formal activity is open. It is worth taking a closer look at some of the prominent views. We can start with philosophy of science. Two important philosophical approaches within the study of science have been *logical positivism*, initially associated with the Vienna Circle, and *falsificationism*, associated with Karl Popper. The Vienna Circle was a group of prominent philosophers and scientists who met in the early 1930s. The project of the Vienna Circle was to develop a philosophical understanding of science that would allow for an expansion of the scientific worldview – particularly into the social sciences and into philosophy itself. That project was immensely successful, because positivism was widely absorbed by scientists and non-scientists interested in increasing the rigor of their work. Interesting conceptual problems, however, caused positivism to become increasingly focused on issues within the philosophy of science, losing sight of the more general project with which the movement began (see Friedman 1999; Richardson 1998).

Logical positivists maintain that the meaning of a scientific theory (and anything else) is exhausted by empirical and logical considerations of what would verify or falsify it. A scientific theory, then, is a condensed summary of possible observations. This is one way in which science can be seen as a formal activity: scientific theories are built up by the logical manipulation of observations (e.g. Ayer 1952 [1936]; Carnap 1952 [1928]), and scientific progress consists in increasing the correctness, number, and range of potential observations that its theories indicate.

For logical positivists, theories develop through a method that transforms individual data points into general statements. The process of creating scientific theories is therefore an inductive one. As a result, positivists tried to develop a logic of science that would make solid the inductive process of moving from individual facts to general claims. For example, scientists might be seen as creating frameworks in which it is possible to uniquely generalize from data (see Box 1.1).

Positivism has immediate problems. First, if meanings are reduced to observations, there are many "synonyms," in the form of theories or statements that look as though they should have very different meanings but do not make different predictions. For example, Copernican astronomy was initially designed to duplicate the (mostly successful) predictions of the earlier Ptolemaic system; in terms of observations, then, the two systems were roughly equivalent, but they clearly meant very different things, since one put the Earth in the center of the universe, and the other had the Earth spinning around the Sun. Second, many apparently meaningful claims are not systematically related to observations, because theories are often too abstract to be immediately cashed out in terms of data. Yet surely abstraction does not render a theory meaningless. Despite these problems and others, the positivist view of meaning taps into deep intuitions, and cannot be entirely dismissed.

Even if one does not believe positivism's ideas about meaning, many people are attracted to the strict relationship that it posits between theories and observations. Even if theories are not mere summaries of observations, they should be absolutely supported by them. The justification we have for believing a scientific theory is based on that theory's solid connection

Box 1.1   The problem of induction

Among the asides inserted into the next few chapters are a number of versions of the "problem of induction." These are valuable background for a number of issues in Science and Technology Studies (STS). At least as stated here, these are theoretical problems that only occasionally become practical ones in scientific and technical contexts. While they could be paralyzing in principle, in practice they do not come up. One aspect of their importance, then, is in finding out how scientists and engineers contain these problems, and when they fail at that, how they deal with them.

The *problem of induction* arose with David Hume's general questions about evidence in the eighteenth century. Unlike classical skeptics, Hume was interested not in challenging particular patterns of argument, but in showing the fallibility of arguments from experience in general. In the sense of Hume's problem, induction extends data to cover new cases. To take a standard example, "the sun rises every 24 hours" is a claim supposedly established by induction over many instances, as each passing day has added another data point to the overwhelming evidence for it. Inductive arguments take *n* cases, and extend the pattern to the *n+1*st. But, says Hume, why should we believe this pattern? Could the *n+1*st case be different, no matter how large *n* is? It does no good to appeal to the regularity of nature, because the regularity of nature is at issue. Moreover, as Ludwig Wittgenstein (1958) and Nelson Goodman (1983 [1954]) show, nature could be perfectly regular and we would still have a problem of induction. This is because there are many possible ideas of what it would mean for the *n+1*st case to be the same as the first *n*. *Sameness* is not a fully defined concept.

It is intuitively obvious that the problem of induction is insoluble. It is more difficult to explain why, but Karl Popper, the political philosopher and philosopher of science, makes a straightforward case that it is. The problem is insoluble, according to him, because there is no principle of induction that is true. That is, there is no way of assuredly going from a finite number of cases to a true general statement about all the relevant cases. To see this, we need only look at examples. "The sun rises every 24 hours" is false, says Popper, as formulated and normally understood, because in Polar regions there are days in the year when the sun never rises, and days in the year when it never sets. Even cases taken as examples of straightforward and solid inductive inferences can be shown to be wrong, so why should we be at all confident of more complex cases?

to data. Another view, then, that is more loosely positivist, is that one can by purely logical means make predictions of observations from scientific theories, and that the best theories are ones that make all the right predictions. This view is perhaps best articulated as *falsificationism*, a position developed by (Sir) Karl Popper (e.g. 1963), a philosopher who was once on the edges of the Vienna Circle.

For Popper, the key task of philosophy of science is to provide a demarcation criterion, a rule that would allow a line to be drawn between science and non-science. This he finds in a simple idea: genuine scientific theories are falsifiable, making risky predictions. The scientific attitude demands that if a theory's prediction is falsified the theory itself is to be treated as false. Pseudo-sciences, among which Popper includes Marxism and Freudianism, are insulated from criticism, able to explain and incorporate any fact. They do not make any firm predictions, but are capable of explaining, or explaining away, anything that comes up.

This is a second way in which science might be seen as a formal activity. According to Popper, scientific theories are imaginative creations, and there is no method for creating them. They are free-floating, their meaning not tied to observations as for the positivists. However, there *is* a strict method for evaluating them. Any theory that fails to make risky predictions is ruled unscientific, and any theory that makes failed predictions is ruled false. A theory that makes good predictions is provisionally accepted – until new evidence comes along. Popper's scientist is first and foremost skeptical, unwilling to accept anything as proven, and willing to throw away anything that runs afoul of the evidence. On this view, progress is probably best seen as the successive refinement and enlargement of theories to cover increasing data. While science may or may not reach the truth, the process of conjectures and refutations allows it to encompass increasing numbers of facts.

Like the central idea of positivism, falsificationism faces some immediate problems. Scientific theories are generally fairly abstract, and few make hard predictions without adopting a whole host of extra assumptions (e.g. Putnam 1981); so on Popper's view most scientific theories would be unscientific. Also, when theories are used to make incorrect predictions, scientists often – and quite reasonably – look for reasons to explain away the observations or predictions, rather than rejecting the theories. Nonetheless, there is something attractive about the idea that (potential) falsification is the key to solid scientific standing, and so falsificationism, like logical positivism, still has adherents today.

For both positivism and falsificationism, the features of science that make it scientific are formal relations between theories and data, whether through

Box 1.2 The Duhem–Quine thesis

The *Duhem–Quine thesis* is the claim that a theory can never be conclusively tested in isolation: what is tested is an entire framework or a web of beliefs. This means that in principle any scientific theory can be held in the face of apparently contrary evidence. Though neither of them put the claim quite this baldly, Pierre Duhem and W.V.O. Quine, writing in the beginning and middle of the twentieth century respectively, showed us why.

How should one react if some of a theory's predictions are found to be wrong? The answer looks straightforward: the theory has been falsified, and should be abandoned. But that answer is too easy, because theories never make predictions in a vacuum. Instead, they are used, along with many other resources, to make predictions. When a prediction is wrong, the culprit might be the theory. However, it might also be the data that set the stage for the prediction, or additional hypotheses that were brought into play, or measuring equipment used to verify the prediction. The culprit might even lie entirely outside this constellation of resources: some unknown object or process that interferes with observations or affects the prediction.

To put the matter in Quine's terms, theories are parts of webs of belief. When a prediction is wrong, one of the beliefs no longer fits neatly into the web. To smooth things out – to maintain a consistent structure – one can adjust any number of the web's parts. With a radical enough redesign of the web, any part of it can be maintained, and any part jettisoned. One can even abandon rules of logic if one needs to!

When Newton's predictions of the path of the moon failed to match the data he had, he did not abandon his theory of gravity, his laws of motion, or any of the calculating devices he had employed. Instead, he assumed that there was something wrong with the observations, and he fudged his data. While fudging might seem unacceptable, we can appreciate his impulse: in his view, the theory, the laws, and the mathematics were all stronger than the data! Later physicists agreed. The problem lay in the optical assumptions originally used in interpreting the data, and when those were changed Newton's theory made excellent predictions.

Does the Duhem–Quine thesis give us a problem of induction? It shows that multiple resources are used (not all explicitly) to make a prediction, and that it is impossible to isolate for blame only one of those resources when the prediction appears wrong. We might, then, see the Duhem–Quine thesis as posing a problem of deduction, not induction, because it shows that when dealing with the real world, many things can confound neat logical deductions.

the rational construction of theoretical edifices on top of empirical data or the rational dismissal of theories on the basis of empirical data. There are analogous views about mathematics; indeed, formalist pictures of science probably depend on stereotypes of mathematics as a logical or mathematical activity.

But there are other features of the popular snapshot of science. These formal relations between theories and data can be difficult to reconcile with an even more fundamental intuition about science: Whatever else it does, science progresses toward truth, and accumulates truths as it goes. We can call this intuition *realism*, the name that philosophers have given to the claim that many or most scientific theories are approximately true.

First, progress. One cannot but be struck by the increases in precision of scientific predictions, the increases in scope of scientific knowledge, and the increases in technical ability that stem from scientific progress. Even in a field as established as astronomy, calculations of the dates and times of astronomical events continue to become more precise. Sometimes this precision stems from better data, sometimes from better understandings of the causes of those events, and sometimes from connecting different pieces of knowledge. And occasionally, the increased precision allows for new technical ability or theoretical advances.

Second, truths. According to realist intuitions, there is no way to understand the increase in predictive power of science, and the technical ability that flows from that predictive power, except in terms of an increase of truth. That is, science can do more when its theories are better approximations of the truth, and when it has more approximately true theories. For the realist, science does not merely construct convenient theoretical descriptions of data, or merely discard falsified theories: When it constructs theories or other claims, those generally and eventually approach the truth. When it discards falsified theories, it does so in favor of theories that better approach the truth.

Real progress, though, has to be built on more or less systematic methods. Otherwise, there would only be occasional gains, stemming from chance or genius. If science accumulates truths, it does so on a rational basis, not through luck. Thus, realists are generally committed to something like formal relations between data and theories.

Turning from philosophy of science, and from issues of data, evidence, and truth, we see a social aspect to the standard picture of science. Scientists are distinguished by their even-handed attitude toward theories, data, and each other. Robert Merton's *functionalist* view, discussed in Chapter 3, dominated discussions of the sociology of science through the 1960s. Merton argued that science served a social function, providing certified knowledge. That function structures norms of scientific behavior, those

Box 1.3   Underdetermination

Scientists choose the best account of data from among competing hypotheses. This choice can never be logically conclusive, because for every explanation there are in principle an indefinitely large number of others that are exactly empirically equivalent. Theories are *underdetermined* by the empirical evidence. This is easy to see through an analogy.

Imagine that our data is the collection of points in the graph on the left (Figure 1.1). The hypothesis that we create to "explain" this data is some line of best fit. But what line of best fit? The graph on the right shows two competing lines that both fit the data perfectly.

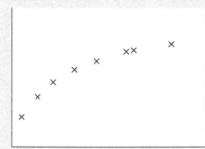 

Clearly there are infinitely many more lines of perfect fit. We can do further testing and eliminate some, but there will always be infinitely many more. We can apply criteria like simplicity and elegance to eliminate some of them, but such criteria take us straight back to the first problem of induction: how do we know that nature is simple and elegant, and why should we assume that our ideas of simplicity and elegance are the same as nature's?

When scientists choose the best theory, then, they choose the best theory from among those that have been seriously considered. There is little reason to believe that the best theory so far considered, out of the infinite numbers of empirically adequate explanations, will be the true one. In fact, if there are an infinite number of potential explanations, we could reasonably assign to each one a probability of zero.

The status of underdetermination has been hotly debated in philosophy of science. Because of the underdetermination argument, some philosophers (positivists and their intellectual descendants) argue that scientific theories should be thought of as instruments for explaining and predicting, not as true or realistic representations (e.g. van Fraassen 1980). Realist philosophers, however, argue that there is no way of understanding the successes of science without accepting that in at least some circumstances evaluation of the evidence leads to approximately true theories (e.g. Boyd 1984; see Box 6.2).

norms that tend to promote the accumulation of certified knowledge. For Merton, science is a well-regulated activity, steadily adding to the store of knowledge.

On Merton's view, there is nothing particularly "scientific" about the people who do science. Rather, science's social structure rewards behavior that, in general, promotes the growth of knowledge; in principle it also penalizes behavior that retards the growth of knowledge. A number of other thinkers hold that position, such as Popper (1963) and Michael Polanyi (1962), who both support an individualist, republican ideal of science, for its ability to progress.

Common to all of these views is the idea that standards or norms are the source of science's success and authority. For positivists, the key is that theories can be no more or less than the logical representation of data. For falsificationists, scientists are held to a standard on which they have to discard theories in the face of opposing data. For realists, good methods form the basis of scientific progress. For functionalists, the norms are the rules governing scientific behavior and attitudes. All of these standards or norms are attempts to define what it is to be scientific. They provide ideals that actual scientific episodes can live up to or not, standards to judge between good and bad science. Therefore, the view of science we have seen so far is not merely an abstraction from science, but is importantly a view of ideal science.

A View of Technology

Where is technology in all of this? Technology has tended to occupy a secondary role, for a simple reason: it is often thought, in both popular and academic accounts, that technology is the relatively straightforward application of science. We can imagine a linear model of innovation, from basic science through applied science to development and production. Technologists identify needs, problems, or opportunities, and creatively combine pieces of knowledge to address them. Technology combines the scientific method with a practically minded creativity.

As such, the interesting questions about technology are about its effects: Does technology determine social relations? Is technology humanizing or dehumanizing? Does technology promote or inhibit freedom? Do science's current applications in technologies serve broad public goals? These are important questions, but as they take technology as a finished product they are normally divorced from studies of the creation of particular technologies.

If technology is applied science then it is limited by the limits of scientific knowledge. On the common view, then, science plays a central role in determining the shape of technology. There is another form of determinism that often arises in discussions of technology, though one that has been more recognized as controversial. A number of writers have argued that the state of technology is the most important cause of social structures, because technology enables most human action. People act in the context of available technology, and therefore people's relations among themselves can only be understood in the context of technology. While this sort of claim is often challenged – by people who insist on the priority of the social world over the material one – it has helped to focus debate almost exclusively on the *effects* of technology.

Lewis Mumford (1934, 1967) established an influential line of thinking about technology. According to Mumford, technology comes in two varieties. *Polytechnics* are "life-oriented," integrated with broad human needs and potentials. Polytechnics produce small-scale and versatile tools, useful for pursuing many human goals. *Monotechnics* produce "mega machines" that can increase power dramatically, but by regimenting and dehumanizing. A modern factory can produce extraordinary material goods, but only if workers are disciplined to participate in the working of the machine. This distinction continues to be a valuable resource for analysts and critics of technology (see, e.g., Franklin 1990, Winner 1986).

In his widely read essay "The Question Concerning Technology" (1977 [1954]), Martin Heidegger develops a similar position. For Heidegger, distinctively modern technology is the application of science in the service of power; this is an objectifying process. In contrast to the craft tradition that produced individualized things, modern technology creates resources, objects made to be used. From the point of view of modern technology, the world consists of resources to be turned into new resources. A technological worldview thus produces a thorough disenchantment of the world.

Through all of this thinking, technology is viewed as simply applied science. For both Mumford and Heidegger modern technology is shaped by its scientific rationality. Even the pragmatist philosopher John Dewey (e.g. 1929), who argues that all rational thought is instrumental, sees science as theoretical technology (using the word in a highly abstract sense) and technology (in the ordinary sense) as applied science. Interestingly, the view that technology is applied science tends toward a form of technological determinism. For example, Jacques Ellul (1964) defines *technique* as "the totality of methods rationally arrived at and having absolute efficiency (for a given stage of development)" (quoted in Mitcham 1994: 308). A society that has accepted modern technology finds itself on a path of increasing efficiency, allowing technique to enter more and more domains. The view that a formal relation between theories and data lies at the core of science informs not only our picture of science, but of technology.

Concerns about technology have been the source of many of the movements critical of science. After the US use of nuclear weapons on Hiroshima and Nagasaki in World War II, some scientists and engineers who had been involved in developing the weapons began *The Bulletin of the Atomic Scientists*, a magazine alerting its readers about major dangers stemming from the military and industrial technologies. Starting in 1955, the Pugwash Conferences on Science and World Affairs responded to the threat of nuclear war, as the United States and the Soviet Union armed themselves with nuclear weapons.

Science and the technologies to which it contributes often result in very unevenly distributed benefits, costs, and risks. Organizations like the Union of Concerned Scientists, and Science for the People, recognized this uneven distribution. Altogether, the different groups that made up the Radical Science Movement engaged in a critique of the idea of progress, with technological progress as their main target (Cutliffe 2000).

Parallel to this in the academy, "Science, Technology and Society" became, starting in the 1970s, the label for a diverse group united by progressive goals and an interest in science and technology as problematic social institutions. For researchers on Science, Technology and Society the project of understanding the social nature of science has generally been seen as continuous with the project of promoting a socially responsible science (e.g. Ravetz 1971; Spiegel-Rösing and Price 1977; Cutliffe 2000). The key issues for Science, Technology and Society are about reform, about promoting disinterested science, and about technologies that benefit the widest populations. How can sound technical decisions be made democratically (Laird 1993)? Can and should innovation be democratically controlled (Sclove 1995)? To what extent, and how, can technologies be treated as political entities (Winner 1986)? Given that researchers, knowledge, and tools flow back and forth between academia and industry, how can we safeguard pure science (Dickson 1988; Slaughter and Leslie 1997)? This is the other "STS," which has played a major role in Science and Technology Studies, the former being both an antecedent of and now a part of the latter.

A Preview of Science and Technology Studies

Science and Technology Studies (STS) starts from an assumption that science and technology are thoroughly social activities. They are social in that scientists and engineers are always members of communities, trained into the practices of those communities and necessarily working within them. These communities set standards for inquiry and evaluate knowledge claims. There is no abstract and logical scientific method apart from evolving community norms. In addition, science and technology are arenas in which rhetorical work is crucial, because scientists and engineers are always in the position of having to convince their peers and others of the value of their favorite ideas and plans – they are constantly engaged in struggles to gain resources and to promote their views. The actors in science and technology are also not mere logical operators, but instead have investments in skills, prestige, knowledge, and specific theories and practices. Even conflicts in a wider society may be mirrored by and connected to conflicts within science and technology; for example, splits along gender, race, class, and national lines can occur both within science and in the relations between scientists and non-scientists.

STS takes a variety of anti-essentialist positions with respect to science and technology. Neither science nor technology is a natural kind, having simple properties that define it once and for all. The sources of knowledge and artifacts are complex and various: there is no privileged scientific method that can translate nature into knowledge, and no technological method that can translate knowledge into artifacts. In addition, the interpretations of knowledge and artifacts are complex and various: claims, theories, facts, and objects may have very different meanings to different audiences.

For STS, then, science and technology are active processes, and should be studied as such. The field investigates how scientific knowledge and technological artifacts are *constructed*. Knowledge and artifacts are human products, and marked by the circumstances of their production. In their most crude forms, claims about the social construction of knowledge leave no role for the material world to play in the making of knowledge about it. Almost all work in STS is more subtle than that, exploring instead the ways in which the material world is used by researchers in the production of knowledge. STS pays attention to the ways in which scientists and engineers attempt to construct stable structures and networks, often drawing together into one account the variety of resources used in making those structures and networks. So a central premise of STS is that scientists and engineers *use* the material world in their work; it is not merely translated into knowledge and objects by a mechanical process.

Clearly, STS tends to reject many of the elements of the common view of science. How and in what respects are the topics of the rest of this book.

2

The Kuhnian Revolution

Thomas Kuhn's *The Structure of Scientific Revolutions* (1970, first published in 1962) challenged the dominant popular and philosophical pictures of the history of science. Rejecting the formalist view with its normative stance, Kuhn focused on the activities of and around scientific research: in his work science is merely what scientists do. Rejecting steady progress, he argued that there have been periods of normal science punctuated by revolutions. Kuhn's innovations were in part an ingenious reworking of portions of the standard pictures of science, informed by rationalist emphases on the power of ideas, by positivist views on the nature and meaning of theories, and by Ludwig Wittgenstein's ideas about forms of life and about perception. The result was novel, and had an enormous impact.

One of the targets of *The Structure of Scientific Revolutions* is what is known (since Butterfield 1931) as "Whig history," history that attempts to construct the past as a series of steps toward (and occasionally away from) present views. Especially in the history of science there is a temptation to see the past through the lens of the present, to see moves in the direction of what we now believe to be the truth as more rational, more natural, and less needing of causal explanation than opposition to what we now believe. But since events must follow their causes, a sequence of events in the history of science cannot be explained teleologically, simply by the fact that they represent progress. Whig history is one of the common buttresses of too-simple progressivism in the history of science, and its removal makes room for explanations that include more irregular changes.

According to Kuhn, *normal science* is the science done when members of a field share a recognition of key past achievements in their field, beliefs about which theories are right, an understanding of the important problems of the field, and methods for solving those problems. In Kuhn's terminology, scientists doing normal science share a *paradigm*. The term, originally referring to a grammatical model or pattern, draws particular attention to

Box 2.1   The modernity of science

Many commentators on science have felt that it is a particularly modern institution. By this they generally mean that it is exceptionally rational, or exceptionally free of local contexts. While science's exceptionality in either of these senses is contentious, there is a straightforward sense in which science is, and always has been, modern. As Derek de Solla Price (1986 [1963]) has pointed out, science has grown rapidly over the past three hundred years. In fact, by any of a number of indicators, science's growth has been steadily exponential. Science's share of the US gross national product has doubled every 20 years. The cumulative number of scientific journals founded has doubled every 15 years, as has the membership in scientific institutes, and the number of people with scientific or technical degrees. The numbers of articles in many sub-fields have doubled every 10 years. These patterns cannot continue indefinitely – and in fact have not continued since Price did his analysis.

A feature of this extremely rapid growth is that between 80 and 90 percent of all the scientists who have ever lived are alive now. For a senior scientist, between 80 and 90 percent of all the scientific articles ever written were written during his or her lifetime. For working scientists the distant past of their fields is almost entirely irrelevant to their current research, because the past is buried under masses of more recent accomplishments. Citation patterns show, as one would expect, that older research is considered less relevant than more recent research, perhaps having been superseded or simply left aside. For Price, a "research front" in a field at some time can be represented by the network of articles that are frequently cited. The front continually picks up new articles and drops old ones, as it establishes new problems, techniques, and solutions. Whether or not there are paradigms as Kuhn sees them, science pays most attention to current work, and little to its past. Science is modern in the sense of having a present-centered outlook, leaving its past to historians.

Rapid growth also gives science the impression of youth. At any time, a disproportionate number of scientists are young, having recently entered their fields. This creates the impression that science is for the young, even though individual scientists may make as many contributions in middle age as in youth (Wray 2003).

a scientific achievement that serves as an example for others to follow. Kuhn also assumes that such achievements provide theoretical and methodological tools for further research. Once they were established, Newton's mechanics, Lavoisier's chemistry, and Mendel's genetics each structured research in their respective fields, providing theoretical frameworks for and models of successful research.

Although it is tempting to see it as a period of stasis, normal science is better viewed as a period in which research is well structured. The theoretical side of a paradigm serves as a *worldview*, providing categories and frameworks into which to slot phenomena. The practical side of a paradigm serves as a *form of life*, providing patterns of behavior or frameworks for action. For example, Lavoisier's ideas about elements and the conservation of mass formed frameworks within which later chemists generated further ideas. The importance he attached to measurement instruments, and the balance in particular, shaped the work practices of chemistry. Within paradigms research goes on, often with tremendous creativity – though always embedded in firm conceptual and social backdrops.

Kuhn talks of normal science as *puzzle-solving*, because problems are to be solved within the terms of the paradigm: failure to solve a problem usually reflects badly on the researcher, rather than on the theories or methods of the paradigm. With respect to a paradigm, an unsolved problem is simply an anomaly, fodder for future researchers. In periods of normal science the paradigm is not open to serious question. This is because the natural sciences, on Kuhn's view, are particularly successful at socializing practitioners. Science students are taught from textbooks that present standardized views of fields and their histories; they have lengthy periods of training and apprenticeship; and during their training they are generally asked to solve well-understood and well-structured problems, often with well-known answers.

Nothing good lasts forever, and that includes normal science. Because paradigms can only ever be partial representations and partial ways of dealing with a subject matter, anomalies accumulate, and may eventually start to take on the character of real problems, rather than mere puzzles. Real problems cause discomfort and unease with the terms of the paradigm, and this allows scientists to consider changes and alternatives to the framework; Kuhn terms this a period of *crisis*. If an alternative is created that solves some of the central unsolved problems, then some scientists, particularly younger scientists who have not yet been fully indoctrinated into the beliefs and practices or way of life of the older paradigm, will adopt the alternative. Eventually, as older and conservative scientists become marginalized, a robust alternative may become a paradigm itself, structuring a new period of normal science.

Box 2.2   Foundationalism

Foundationalism is the thesis that knowledge can be traced back to firm foundations. Typically those foundations are seen as a combination of sensory impressions and rational principles, which then support an edifice of higher-order beliefs. The central metaphor of foundationalism, of a building firmly planted in the ground, is an attractive one. If we ask why we hold some belief, the reasons we give come in the form of another set of beliefs. We can continue asking why we hold *these* beliefs, and so on. Like bricks, each belief is supported by more beneath it (there is a problem here of the nature of the mortar that holds the bricks together, but we will ignore that). Clearly, the wall of bricks cannot continue downward forever; we do not support our knowledge with an infinite chain of beliefs. But what lies at the foundation?

The most plausible candidates for empirical foundations are sense experiences. But how can these ever be combined to support the complex generalizations that form our knowledge? We might think of sense experiences, and especially their simplest components, as like individual data points. Here we have the earlier problems of induction all over again: as we have seen, a finite collection of data points cannot determine which generalizations to believe.

Worse, even beliefs about sense impressions are not perfectly secure. Much of the discussion around Kuhn's *The Structure of Scientific Revolutions* (1970 [1962]) has focused on his claim that scientific revolutions change what scientists observe (Box 2.3). Even if Kuhn's emphasis is wrong, it is clear that we often doubt what we see or hear, and reinterpret it in terms of what we know. The problem becomes more obvious, as the discussion of the Duhem–Quine thesis (Box 1.2) shows, if we imagine the foundations to be already-ordered collections of sense impressions.

On the one hand, then, we cannot locate plausible foundations for the many complex generalizations that form our knowledge. On the other hand, nothing that might count as a foundation is perfectly secure. We are best off to abandon, then, the metaphor of solid foundations on which our knowledge sits.

According to Kuhn, it is in periods of normal science that we can most easily talk about progress, because scientists have little difficulty recognizing each other's achievements. Revolutions, however, are not progressive, because they both build and destroy. Some or all of the research structured by the pre-revolutionary paradigm will fail to make sense under the new regime; in fact Kuhn even claims that theories belonging to different paradigms are *incommensurable* – lacking a common measure – because people working in different paradigms see the world differently, and because the meanings of theoretical terms change with revolutions (a view derived in part from positivist notions of meaning). The non-progressiveness of revolutions and the incommensurability of paradigms are two closely related features of the Kuhnian account that have caused many commentators the most difficulty.

If Kuhn is right, science does not straightforwardly accumulate know-ledge, but instead moves from one more or less adequate paradigm to another. This is the most radical implication found in *The Structure of Scientific Revolutions*: Science does not track the truth, but creates different partial views that can be considered to contain truth only by people who hold those views!

Kuhn's claim that theories within paradigms are incommensurable has a number of different roots. One of those roots lies in the positivist picture of meaning, on which the meanings of theoretical terms are related to observations they imply. Kuhn adopts the idea that the meanings of theoretical terms depend upon the constellation of claims in which they are embedded. A change of paradigms should result in widespread changes in the meanings of key terms. If this is true, then none of the key terms from one paradigm would map neatly onto those of another, preventing a common measure, or even full communication.

Secondly, in *The Structure of Scientific Revolutions*, Kuhn takes the notion of indoctrination quite seriously, going so far as to claim that paradigms even shape observations. People working within different paradigms see things differently. Borrowing from the work of N. R. Hanson (1958), Kuhn argues there is no such thing, at least in normal circumstances, as raw observation. Instead, observation comes interpreted: we do not see dots and lines in our visual fields, but instead see more or less recognizable objects and patterns. Thus observation is guided by concepts and ideas. This claim has become known as the *theory-dependence of observation*. The theory-dependence of observation is easily linked to Kuhn's historical picture, because during revolutions people stop seeing one way, and start seeing another way, guided by the new paradigm.

Finally, one of the roots of Kuhn's claims about incommensurability is his experience as an historian that it is difficult to make sense of past sci-entists' problems, concepts, and methods. Past research can be opaque, and aspects of it can seem bizarre. It might even be said that if people find it too easy to understand very old research in present terms they are probably doing some interpretive violence to that research – Isaac Newton's physics looks strikingly modern when rewritten for today's textbooks, but looks much less so in its originally published form, and even less so when the connections between it and Newton's religious and alchemical research are drawn (e.g. Dobbs and Jacob 1995). Kuhn says that "In a sense that I am unable to explicate further, the proponents of competing paradigms practice their trades in different worlds" (1970 [1962]: 150).

The case for semantic incommensurability has attracted a considerable amount of attention, mostly negative. Meanings of terms do change, but they probably do not change so much and so systematically that claims in which they are used cannot typically be compared. Most of the philosophers, linguists, and others who have studied this issue have come to the conclusion that claims for semantic incommensurability cannot be sustained, or even that it is impossible (Davidson 1974) to make sense of such radical change in meaning (see Bird 2000 for an overview).

This leaves the historical justification for incommensurability. That problems, concepts, and methods change is uncontroversial. But the difficulties that these create for interpreting past episodes in science can be overcome – the very fact that historical research can challenge present-centered interpretations shows the limits of incommensurability.

Claims of radical incommensurability appear to fail. In fact, Kuhn quickly distanced himself from the strongest readings of his claims. Already by 1965 he insisted that he meant by "incommensurability" only "incomplete communication" or "difficulty of translation," sometimes leading to "communication breakdown" (Kuhn 1970a). Still, on these more modest readings incommensurability is an important phenomenon: even when dealing with the same subject matter, scientists (among others) can fail to communicate.

If there is no radical incommensurability, then there is no radical division between paradigms, either. Paradigms must be linked by enough continuity of concepts and practices to allow communication. This may even be a methodological or theoretical point: complete ruptures in ideas or practices are inexplicable (Barnes 1982). When historians want to explain an innovation, they do so in terms of a reworking of available resources. Every new idea, practice, and object has its sources; to assume otherwise is to invoke something akin to magic. Thus many historians of science have challenged Kuhn's paradigms by showing the continuity from one putative paradigm to the next.

For example, instruments, theories, and experiments change at different times. In a detailed study of particle detectors in physics, Peter Galison (1997) shows that new detectors are initially used for the same types of experiments and observations as their immediate predecessors had been, and fit into the same theoretical contexts. Similarly, when theories change, there is no immediate change in either experiments or instruments. Discontinuity in one realm, then, is at least generally bounded by continuity in others. Science gains strength, an *ad hoc* unity, from the fact that its key components rarely change together. Science maintains stability through change by being dis-unified, like a thread as described by Wittgenstein (1958): "the strength of the thread does not reside in the fact that some one fibre runs through its whole length, but in the overlapping of many fibres." If this is right then the image of complete breaks between periods is misleading.

Box 2.3   The theory-dependence of observation

Do people's beliefs shape their observations? Psychologists have long studied this question, showing how people's interpretations of images are affected by what they expect those images to show. Hanson and Kuhn took the psychological results to be important for understanding how science works. Scientific observations, they claim, are theory-dependent.

For the most part, philosophers, psychologists, and cognitive scientists agree that observations can be shaped by what people believe. There are substantial disagreements, though, about how important this is for understanding science. For example, a prominent debate about visual illusions and the extent to which the background beliefs that make them illusions are plastic (e.g. Churchland 1988; Fodor 1988) has been sidelined by a broader interpretation of "observation." Scientific observation has been and is rarely equivalent to brute perception, experienced by an isolated individual (Daston 2008). Much scientific data is collected by machine, and then is organized by scientists to display phenomena publicly (Bogen and Woodward 1992). If that organization amounts to observation, then it is straightforward that observation is theory-dependent.

Theory and practice dependence is broader even than that: scientists attend to objects and processes that background beliefs suggest are worth look-ing at, they design experiments around theoretically inspired questions, they remember relevance and communicate relevant information, where relevance depends on established practices and shared theoretical views (Brewer and Lambert 2001).

Incommensurability: Communicating Among Social Worlds

Claims about the incommensurability of scientific paradigms raise general questions about the extent to which people across boundaries can communicate.

In some sense it is trivial that disciplines (or smaller units, like specialties) are incommensurable. The work done by a molecular biologist is not obviously interesting or comprehensible to an evolutionary ecologist or a neuropathologist, although with some translation it can sometimes become so. The meaning of terms, ideas, and actions is connected to the cultures and practices from which they stem. Disciplines are "epistemic cultures" that may have completely different orientations to their objects, social units of knowledge production, and patterns of interaction (Knorr Cetina 1999). However, people from different areas interact, and as a result science gains a degree of unity. We might ask, then, how interactions are made to work.

Simplified languages allow parties to trade goods and services without concern for the integrity of local cultures and practices. A *trading zone* (Galison 1997) is an area in which scientific and/or technical practices can fruitfully interact via these simplified languages or *pidgins*, without requiring full assimilation. Trading zones can develop at the contact points of specialties, around the transfer of valuable goods from one to another. In trading zones, collaborations can be successful even if the cultures and practices that are brought together do not agree on problems or definitions.

The trading zone concept is flexible, perhaps overly so. We might look at almost any communication as taking place in a trading zone and demanding some pidgin-like language. For example, Richard Feynman's diagrams of particle interactions, which later became known as Feynman diagrams, were successful in part because they were simple and could be interpreted in various ways (Kaiser 2005). They were widely spread during the 1950s by visiting postdoctoral fellows and researchers. But different schools, working with different theoretical frameworks, picked them up, adapted them, and developed local styles of using them. Despite their variety, they remained important ways of communicating among physicists, and also tools that were productive of theoretical problems and insights. It would seem to stretch the "trading zone" concept to say that Feynman diagrams were parts of pidgins needed for theoretical physicists to talk to each other, yet that is what they look like.

A different, but equally flexible, concept for understanding communication across barriers is the idea of *boundary objects* (Star and Griesemer 1989). In a historical case study of interactions in Berkeley's Museum of Vertebrate Zoology, Susan Leigh Star and James Griesemer focus on objects, rather than languages. The different social worlds of amateur collectors, professional scientists, philanthropists, and administrators had very different visions of the museum, its goals, and the important work to be done. These differences resulted in incommensurabilities among groups. However, objects can form bridges across boundaries, if they can serve as a focus of attention in different social worlds, and are robust enough to maintain their identities in those different worlds.

Standardized records were among the key boundary objects that held together these different social worlds. Records of the specimens had different meanings for the different groups of actors, but each group could contribute to and use those records. The practices of each group could continue intact, but the groups interacted via record keeping. Boundary objects, then, allow for a certain amount of coordination of actions without large measures of translation.

The boundary object concept has been picked up and used in an enormous number of ways. Even within the article in which they introduce the concept, Star and Griesemer present a number of different examples of boundary objects, including the zoology museum itself, the different animal species in the museum's scope, the state of California, and standardized records of specimens.

The concept has been applied very widely in STS. To take just a few examples: Sketches and drawings can allow engineers in different parts of design and production processes to communicate across boundaries (Henderson 1991). Parameterizations of climate models, the filling in of variables to bring those models in line with the world's weather, connect field meteorologists and simulation modelers (Sundberg 2007). In the early twentieth century breeds of rabbits and poultry connected fanciers to geneticists and commercial breeders (Marie 2008).

Why are there so many different boundary objects? The number and variety suggest that, despite some incommensurability across social boundaries, there is considerable coordination and probably even some level of communication. For example, in multidisciplinary research a considerable amount of communication is achieved via straightforward translation (Duncker 2001). Researchers come to understand what their colleagues in other disciplines know, and translate what they have to say into a language that those colleagues can understand. Simultaneously, they listen to what other people have to say and read what other people write, attuned to differences in knowledge, assumptions, and focus. Concepts like pidgins, trading zones, and boundary objects, while they might be useful in particular situations, may overstate difficulties in communication. Incommensurability as it is found in many practices may not always be a very serious barrier.

The divisions of the sciences result in disunity (see Dupré 1993; Galison and Stump 1996). A disunified science requires communication, perhaps in trading zones or direct translation, or coordination, perhaps via boundary objects, so that its many fibers are in fact twisted around each other. Even while disunified, though, science hangs together and has some stability. How it does so remains an issue that merits investigation.

Conclusion: Some Impacts

The Structure of Scientific Revolutions had an immediate impact. The word "paradigm," referring to a way things are done or seen, came into common usage largely because of Kuhn. Even from the short description above it is clear that the book represents a challenge to earlier important beliefs about science.

Against the views of science with which we started, *The Structure of Scientific Revolutions* argues that scientific communities are importantly organized around ideas and practices, not around ideals of behavior. And, they are organized from the bottom up, not, as functionalism would have it, to serve an overarching goal. Against positivism, Kuhn argued that changes in theories are not driven by data but by changes of vision. In fact, if worldviews are essentially theories then data is subordinate to theory, rather than the other way around. Against falsificationism, Kuhn argued that anomalies are typically set aside, that only during revolutions are they used as a justification to reject a theory. And against all of these he argued that on the largest scales the history of science should not be told as a story of uninterrupted progress, but only change.

Because Kuhn's version of science violated almost everybody's ideas of the rationality and progress of science, *The Structure of Scientific Revolutions* was sometimes read as claiming that science is fundamentally irrational, or describing science as "mob rule." In retrospect it is difficult to find much irrationalism there, and possible to see the book as somewhat conservative – perhaps not only intellectually conservative but politically conservative (Fuller 2000). More important, perhaps, is the widespread perception that by examining history Kuhn firmly refuted the standard view of science.

Whether or not that is true, Kuhn started people thinking about science in very different terms. The success of the book created a space for thinking about the practices of science in local terms, rather than in terms of their contribution to progress, or their exemplification of ideals. Though few of Kuhn's specific ideas have survived fully intact, *The Structure of Scientific Revolutions* has profoundly affected subsequent thinking in the study of science and technology.

3

Questioning Functionalism in the Sociology of Science

Structural-functionalism

Robert Merton's statement, "The institutional goal of science is the extension of certified knowledge" (1973: 270), is the supporting idea behind his thinking on science. His structural-functionalist view assumes that society as a whole can be analyzed in terms of overarching institutions such as religion, government, and science. Each institution, when working well, serves a necessary function, contributing to the stability and flourishing of society. To work well, these institutions must have the appropriate structure. Merton treats science, therefore, as a roughly unified and singular institution, the function of which is to provide certified knowledge. The work of the sociologist is primarily to study how its social structure does and does not support its function. Merton is the most prominent of functionalist sociologists of science, and so his work is the main focus of this chapter, to the neglect of such sociologists as Joseph Ben-David (1991) and John Ziman (1984), and sociologically minded philosophers like David Hull (1988).

The key to Merton's theory of the social structure of science lies in the ethos of science, the norms of behavior that guide appropriate scientific practice. Merton's norms are institutional imperatives, in that rewards are given to community members who follow them, and sanctions are applied to those who violate them. Most important in this ethos are the four norms first described in 1942: universalism, communism, disinterestedness, and organized skepticism.

Universalism requires that the criteria used to evaluate a claim not depend upon the identity of the person making the claim: "race, nationality, religion, class, and personal qualities are irrelevant" (Merton 1973: 270). This should stem from the supposed impersonality of scientific laws; they are either true or false, regardless of their proponents and their provenance. How does the norm of universalism apply in practice? We might look to science's many peer review systems. For example, most scientific journals accept articles for publication based on evaluations by experts. And in most fields, those experts are not told the identity of the authors whose articles they are reviewing. Although not being told the author's name does not guarantee his or her anonymity – because in many fields a well-connected reviewer can guess the identity of an author from the content of the article – it supports universalism nonetheless, both in practice and as an ideal.

Communism states that scientific knowledge – the central product of science – is commonly owned. Originators of ideas can claim recognition for their creativity, but cannot dictate how or by whom those ideas are to be used. Results should be publicized, so that they can be used as widely as possible. This serves the ends of science, because it allows researchers access to many more findings than they could hope to create on their own. According to Merton, communism not only promotes the goals of science but reflects the fact that science is a social activity, or that scientific achievements are collectively produced. Even scientific discoveries by isolated individuals arise as a result of much earlier research.

Disinterestedness is a form of integrity, demanding that scientists disengage their interests from their actions and judgments. They are expected to report results fully, no matter what theory those results support. Disinterestedness should rule out fraud, such as reporting fabricated data, because fraudulent behavior typically represents the intrusion of interests. And indeed, Merton believes that fraud is rare in science.

Organized skepticism is the tendency for the community to disbelieve new ideas until they have been well established. Organized skepticism operates at two levels. New claims are often greeted by arrays of public challenges. For example, even an audience favorably disposed to its claims may fiercely question a presentation at a conference. In addition, scientists may privately reserve judgment on new claims, employing an internalized version of the norm.

In addition to these "moral" norms there are "cognitive" norms concerning rules of evidence, the structure of theories, and so on. Because Merton drew a firm distinction between social and technical domains, cognitive norms are not a matter for his sociology of science to investigate. In general, Merton's sociology does not make substantial claims about the intellectual content of science.

Institutional norms work in combination with rewards and sanctions, in contexts in which community members are socialized to respond to those rewards and sanctions. Rewards in the scientific community are almost entirely honorific. As Merton identifies them, the highest rewards come via eponymy: *Darwinian biology*, the *Copernican system*, *Planck's constant*, and *Halley's comet* all recognize enormous achievements. Other forms of honorific reward are prizes and historical recognition; the most ordinary form of scientific reward is citation of one's work by others, seen as an indication of influence. Sanctions are similarly applied in terms of recognition, as the reputations of scientists who display deviant behavior suffer.

In the 1970s, the Mertonian picture of the ethos of science came under attack, on a variety of instructive grounds. Although there were many criticisms, probably the three most important questions asked were: (1) Is the actual conduct of science governed by Mertonian norms? To be effective, norms of behavior must become part of the culture and institutions of science. In addition, there must be sanctions that can be applied when scientists deviate from the norms; but there is little evidence of strong sanctions for violation of these norms. (2) Are these norms too flexible or vague to perform any analytic or scientific work? (3) Does it make sense to talk of an institutional or overarching goal of something as complex, divided, and evolving as science? These and other questions created a serious challenge to that view, a challenge that helped to push STS toward more local, action-oriented views.

Ethos and Ethics

Social norms establish not only an ethos of science but an ethics of science. Violations of norms are, importantly, ethical lapses. This aspect of Merton's picture has given rise to some interesting attempts to understand and define scientific misconduct, a topic of increasing public interest (Guston 1999a).

On the structural-functionalist view, the public nature of science should mean that deviant behavior is rare. At the same time, deviance is to be expected, as a result of conflicts among norms. In particular, science's reward system is the payment for contributions to communally owned results. However, the pressures of recognition can often create pressures to violate other norms. A disinterested attitude toward one's own data, for example, may go out the window when recognition is importantly at stake, and this may create pressure to fudge results. Fraud and other forms of scientific misconduct occur because of the structures that advance knowledge, not despite them.

Questions of misconduct often run into a problem of differentiating between fraud and error, both of which can stand in the way of progress. The structural-functionalist view explains why fraud is reprehensible, while error is merely undesirable. The difference between them is the difference between the violation of social and cognitive norms (Zuckerman 1977, 1984).

Such models continue to shape discussions of scientific misconduct. The US National Academy of Sciences' primer on research ethics, *On Being a*

Box 3.1   Is fraud common?

There are enormous pressures on scientists to perform, and to establish careers. Yet there are difficulties in replicating experiments, there is an elite system that allows some researchers to be relatively immune from scrutiny, and there is an unwillingness of the scientific community to level accusations of outright fraud (Broad and Wade 1982). It is difficult, then, to know just how common fraud is, but there is reason to suspect that it might be common.

Because of its substantial role in funding scientific research, the US Congress has on several occasions held hearings to address fraud. Prominently, Congressman Albert Gore, Jr. held hearings in 1981 in response to a rash of allegations of fraud at prominent institutions, and Congressman John Dingell held a series of hearings, starting in 1988, that featured "the Baltimore case" (Kevles 1998).

David Baltimore was a Nobel Prize-winning biologist who became entangled in accusations against one of his co-authors on a 1986 publication. The events became "the Baltimore case" because he was the most prominent of the scientific actors, and because he persistently and sometimes pugnaciously defended the accused researcher, Thereza Imanishi-Kari. In 1985, Imanishi-Kari was an immunologist at the Massachusetts Institute of Technology (MIT), under pressure to publish enough research to merit tenure. She collaborated with Baltimore and four other researchers on an experiment on DNA rearrangement, the results of which were published. A postdoctoral researcher in Imanishi-Kari's laboratory, Margot O'Toole, was assigned some follow-up research, but was unable to repeat the original results. O'Toole became convinced that the published data was not the same as the data contained in the laboratory notebooks.

After a falling-out between Imanishi-Kari and O'Toole and a graduate student, Charles Maplethorpe, questions about fraud started working their way up through MIT. Settled in Imanishi-Kari's favor at the university, Maplethorpe alerted National Institutes of Health scientists Ned Feder and Walter W. Stewart to the controversy. Because of an earlier case, Feder and Stewart had become magnets for, and were on their way to becoming advocates of, the investigation of scientific fraud. They brought the case to the attention of Congressman Dingell.

In the US Congress the case became a much larger confrontation. Baltimore defended Imanishi-Kari and attacked the inquiry as a witch-hunt; a number of his scientific colleagues thought his tack unwise, because of the publicity he generated, and because he was increasingly seen as an interested party. Dingell found in Baltimore an opponent who was important enough to be worth taking down, and in O'Toole a convincing witness. Over the course of the hearings, Baltimore's conduct was made to look unprofessional, to the extent that he resigned his position as President of Rockefeller University.

However, Imanishi-Kari was later exonerated, and Baltimore was seen as having taken a courageous stand (Kevles 1998). This raises questions about the nature of any accusation of fraud. At the same time, though, the case reinforces suspicions about the possible commonness of scientific fraud: the pressure to publish was substantial; the experiments were difficult to repeat; whether there had been fraud, or even substantial error, was open to interpretation; and the local scientific investigation was quick, though perhaps correct, to find no evidence of fraud.

Scientist (1995), is a widely circulated booklet containing discussions of different scenarios and principles. Ethical norms, more concrete and nuanced than Merton's, are presented as being in the service of the advancement of knowledge. That is, the resolution of most ethical problems in science typically turns on understanding how to best maintain the scientific enterprise. Functionalism about science, then, can translate more or less directly into ethical advice.

Is the Conduct of Science Governed by Mertonian Norms?

Are the norms of science constant through history and across science? A cursory look at different broad periods suggests that they are not constant, and consideration of different roles that scientists can play shows that norms can be interpreted differently by different actors (e.g. Zabusky and Barley 1997). Are they distinctive to science? Universalism, disinterestedness, and organized skepticism are at some level *professed* norms for many activities in many societies, and may not be statistically more common in science than elsewhere. Disinterestedness, for example, is a version of a norm of rationality, in that it privileges rationality over special interests, but rationality is professed nearly everywhere. People inside and outside of science claim that they generally act rationally. What evidence could show us that science is particularly rational?

What about social versus cognitive norms? As we saw in the last chapter, Kuhn describes the work of normal science as governed by a paradigm, and thus by ideas specific to particular areas of research and times. If this is right, then normal science is shaped by solidarities built around key ideas, not around general behaviors. For example, Kuhn sees scientific education as authoritarian, militating against skepticism in favor of commonly held general beliefs. It seems likely that cognitive norms are more important to scientists' work than are any general moral norms (Barnes and Dolby 1970).

This point can be seen in another criticism of Mertonian norms, put forward by Michael Mulkay (1969), using the example of the furor over the work of Immanuel Velikovsky. In his 1950 book *Worlds in Collision* Velikovsky argued that historical catastrophes, recorded in the Bible and elsewhere, were the result of a near-collision between Earth and a planet-sized object that broke off of Jupiter. The majority of mainstream scientists saw this as sensational pseudo-science. Mulkay uses the case to show one form of deviance from Mertonian norms in science:

> In February, 1950, severe criticisms of Velikovsky's work were published in *Science News Letter* by experts in the fields of astronomy, geology, archaeology, anthropology, and oriental studies. None of these critics had at that time seen *Worlds in Collision*, which was only just going into press. Those denunciations were founded upon popularized versions published, for example, in *Harper's*, *Reader's Digest* and *Collier's*. The author of one of these articles, the astronomer Harlow Shapley, had earlier refused to read the manuscript of Velikovsky's book because Velikovsky's "sensational claims" violated the laws of mechanics. Clearly the "laws of mechanics" here operate as norms, departure from which cannot be tolerated. As a consequence of Velikovsky's non-conformity to these norms Shapley and others felt justified in abrogating the rules of universalism and organized skepticism. They judged the man instead of his work . . . (Mulkay 1969: 32–33)

Scientists violated Mertonian norms in the name of a higher one: claims should be consistent with well-established truths. One could argue that, even on Mertonian terms, violation of the norms in the name of truth makes sense, since those norms are supposed to represent a social structure that aids the discovery of truths. Nonetheless, this type of case shows one way in which moral norms are subservient to cognitive norms.

So far, we have seen that Merton's social norms may not be as important as cognitive norms to understanding the practice of science. But what if we looked at the practice of science and discovered that the opposites of those norms – secrecy, particularism, interestedness, and credulity – were common?

Do scientific communities and their institutions sanction researchers who are, say, secretive about their work? There are, after all, obvious reasons to be secretive. If other researchers learn about one's ideas, methods, or results, they may be in a position to use that information to take the next steps in a program of research, and receive full credit for whatever comes of those steps. Given that science is highly competitive, and given that an increasing amount of science is linked to applications on which there are possible financial stakes (Chapter 17), there are strong incentives to follow through on a research program before letting other researchers know about it. On the structural-functionalist picture, norms exist to counteract local interests such as recognition and monetary gain, so that the larger goal – the growth of knowledge – is served. If Merton is right, we should expect to see violations of norms subject to sanctions.

In a study of scientists working on the Apollo moon project, Ian Mitroff (1974) shows not only that scientists do not apply sanctions, but that they often respect what he calls counter-norms, which are rough opposites of Mertonian ones. Scientists interviewed by Mitroff voiced approval of, for example, interested behavior (1974: 588): "Commitment, even extreme commitment such as bias, has a role to play in science and can serve science well." "Without commitment one wouldn't have the energy, the drive to press forward against extremely difficult odds." "The [emotionally] disinterested scientist is a myth. Even if there were such a being, he probably wouldn't be worth much as a scientist." Mitroff's subjects identified positive value in opposites to each of Merton's norms: Scientific claims are judged in terms of who makes them. Secrecy is valued because it allows scientists to follow through on research programs without worrying about other people doing the same work. Dogmatism allows people to build on others' results without worrying about foundations.

If there are both norms and counter-norms, then the analytical framework of norms does no work. A framework of norms and counter-norms can justify anything, which means that it does not help to understand anything. Moreover, this is not just a methodological problem for theorists, but is also a problem for norm-based actions. When scientists act, norms and counter-norms can give them no guidance and cannot cause them to do anything. The reasons for or causes of actions must lie elsewhere.

Interpretations of Norms

Norms have to be interpreted. This represents a problem for the analyst, but also shows that the force of norms is limited. Let us return to Mulkay's

Box 3.2   Wittgenstein on rules

Wittgenstein's discussion of rules and following rules has been seen as foundational to STS. Although it is complex, the central point can be seen in a short passage. Wittgenstein asks us to imagine a student who has been taught basic arithmetic. We ask this student to write down a series of numbers starting with zero, adding two each time (0, 2, 4, 6, 8, . . .).

> Now we get the pupil to continue . . . beyond 1000 – and he writes 1000, 1004, 1008, 1012. We say to him: "Look what you've done!" – He doesn't understand. We say: "You were meant to add *two*: look how you began the series!" – He answers: "Yes, isn't it right? I thought that was how I was *meant* to do it." – Or suppose he pointed to the series and said: "But I went on in the same way." – It would now be no use to say: "But can't you see . . . ?" – and repeat the old examples and explanations. – In such a case we might say, perhaps: It comes natural to this person to understand our order with our explanations as *we* should understand the order: "Add 2 up to 1000, 4 up to 2000, 6 up to 3000, and so on." (Wittgenstein 1958: Paragraph 185)

Of course this student can be corrected, and can be taught to apply the rule as we would – there is coercion built into such education – but there is always the possibility of future differences of opinion as to the meaning of the rule. In fact, Wittgenstein says, "no course of action could be determined by a rule, because every course of action can be made out to accord with a rule" (Paragraph 201). Rules do not contain the rules for the scope of their own applicability.

Wittgenstein's problem is an extension of Hume's problem of induction. A finite number of examples, with a finite amount of explanation, cannot constrain the next unexamined case. The problem of rule following becomes a usefully different problem because it is in the context of actions, and not just observations.

There are competing interpretations of Wittgenstein's writing on this problem. Some take him as posing a skeptical problem and giving a skeptical solution: people come to agreement about the meaning of rules because of prior socialization, and continuing social pressure (Kripke 1982). Others take him as giving an anti-skeptical solution after showing the absurdity of the skeptical position: hence we need to understand rules not as formulas standing apart from their application, but as constituted by their application (Baker and Hacker 1984). Exactly the same debate has arisen within STS (Bloor 1992; Lynch 1992a, 1992b; Kusch 2004). For our purposes here it is not crucial which of these positions is right, either about the interpretation of Wittgenstein or about rules, because both sides agree that expressions of rules do not determine their applications.

example of the Velikovsky case. The example was originally used to show that scientists violated norms when a higher norm was at stake. Mulkay later noticed that, depending on which parts of the context one attends to, the norms can be interpreted as having been violated or not.

> It could be argued that the kind of qualitative, documentary evidence used by Velikovsky had been shown time and time again to be totally unreliable as a basis for impersonal scientific analysis and that to treat this kind of pseudo-science seriously was to put the whole scientific enterprise in jeopardy. In this way scientists could argue that their response to Velikovsky was an expression of organized skepticism and an attempt to safeguard universalistic criteria of scientific adequacy. (Mulkay 1980: 112)

The problem points to a more general problem about following rules (Box 3.2): Behaviors can be interpreted as following or not following the norms. We can explain almost any scientific episode as one of adherence to Mertonian norms, or as one of the violation of those norms. Thus, in a further way, they are analytically weak.

As in the case of counter-norms, the problem is not just a methodological one. If we as onlookers can interpret the actions of scientists as either in conformity to or in violation of the norms, so can the participants themselves. But that simply means that the norms do not constrain scientists. By creatively selecting contexts, any scientist can use the norms to justify almost any action. And if norms do not represent constraints, then they do no scientific work.

Norms as Resources

Recognizing that norms can be interpreted flexibly suggests that we study not how norms work, but how they are used. That is, in the course of explaining and criticizing actions, scientists invoke norms – such as the Mertonian norms, but in principle an indefinite number of others. For example, because of his refusal to accept the truth of quantum mechanics, Albert Einstein is often seen as becoming conservative as he grew older; being "conservative" clearly violates the norm of disinterestedness (Kaiser 1994). Einstein is so labeled in order to understand how the same person who revolutionized accepted notions of space and time could later reject a theory because it challenged accepted notions of causality: otherwise how could the twentieth century's epitome of the scientific genius make such a mistake? Implicit in the charge, however, is an assumption that Einstein was wrong to reject quantum mechanics, an assumption that quantum mechanics is obviously right, whatever the difficulties that some people have with it. Werner Heisenberg, one of the participants in the debate, discounted Einstein's positions by claiming that they were produced by closed-minded dogmatism and old age. If we believe Heisenberg, we can safely ignore critics of quantum mechanics. How are norms serving as resources in this case? They are being used to help eliminate conflicting views: because Einstein's opposition to quantum mechanics violated norms of conduct, we do not have to pay much attention to his arguments.

Whether a theory stands or falls depends upon the strengths of the arguments put forward for and against it (it also depends upon the theory's usefulness, upon the strengths of the alternatives, and so on). However, it is rarely simple to evaluate important and real theories, and so complex arguments are crucial to science and to scientific beliefs. Norms of behavior can play a role, if they are used to diminish the importance of some arguments and increase the importance of others. Supporters of quantum mechanics are apt to see Einstein as a conservative in his later years. Opponents of quantum mechanics are apt to see him as maintaining a youthful skepticism throughout his life (Fine 1986). Norms are ideals, and like all ideals, they do not apply straightforwardly to concrete cases. People with different interests and different perspectives will apply norms differently.

We are led, then, from seeing norms as constraining actions to seeing them as rhetorical resources. This is one of many parallel shifts of focus in STS, of which we will see more in later chapters. For the most part, these are changes from more structure-centered perspectives to more agent- or action-centered perspectives. This is not to say that there is one simple theoretical maneuver that characterizes STS, but that the field has found some shifts from structure- to action-centered perspectives to be particularly valuable.

Boundary Work

The study of "boundary work" is one approach to seeing norms as resources (Gieryn 1999). When issues of epistemic authority, the authority to make respected claims, arise, people attempt to draw boundaries. To have authority on any contentious issue requires that at least some other people do not have it. The study of boundary work is a localized, historical, or anti-foundational approach to understanding authority (Gieryn 1999). For example, some people might argue that science gets its epistemic authority from its rationality, its connection to nature, or its connection to technology or policy. We can see those connections, though, as products of boundary work: Science is rational because of successful efforts to define it in terms of rationality; science is connected to nature because it has acquired authority to determine what nature is; and scientists connect their work to the benefits

Box 3.3 Cyril Burt, from hero to fraud

Sir Cyril Burt (1883–1971) was one of the most eminent psychologists of the twentieth century, and knighted for his contributions to psychology and to public policy. Burt was known for his strong data and arguments supporting hereditarianism (nature) over environmentalism (nurture) about intelligence. After his death, opponents of hereditarianism pointed out that his findings were curiously consistent over the years. In 1976 in *The Times*, a medical journalist, Oliver Gillie, accused Burt of falsifying data, inventing studies and even co-workers. This public accusation of fraud against one of the discipline's most noted figures posed a challenge to the authority of the psychology itself (Gieryn and Figert 1986).

Early on, his supporters represented Burt as occasionally sloppy, but insisted that there was no evidence of fraud. Burt's work was difficult, they argued, and it was therefore understandable that he made some mistakes. No psychologist's work would be immune from criticism. In addition, Burt was an "impish" character, explaining his invention of colleagues. These responses construed Burt's work as scientific, but science as imperfect. That is, psychologists drew boundaries that accepted minor flaws in science, and thus allowed a flawed character to be one of their own.

In addition to denying or minimizing the accusations, responses by psychologists involved charging Gillie with acting inappropriately. By publishing his accusations in a newspaper, Gillie had subjected Burt's work to a trial not by his peers. The public nature of the IQ controversy raised questions about motives: Were environmentalists trumping up or blowing up the accusations to discredit the strongest piece of evidence against them? Psychologists insisted that there be a *scientific* inquiry into the matter, and endorsed the ongoing research of one of their own, Leslie Hearnshaw, who was working on a biography of Burt. That biography ended up agreeing with the accusers. However, it rescued psychology by banishing Burt, and using the idea that "the truth will out" in science to recover the discipline's authority. Hearnshaw argued that the fraud was the result of personal crisis, especially late Burt's in life, and was the result of his acting in a particularly unscientific manner. Most importantly, he argued that Burt was not a real scientist, but was rather an outsider who sometimes did good scientific work:

> The gifts which made Burt an effective applied psychologist . . . militated against his scientific work. Neither by temperament nor by training was he a scientist. He was overconfident, too much in a hurry, too eager for final results, too ready to adjust and paper-over, to be a good scientist. His work often had the appearance of science, but not always the substance. (Hearnshaw, quoted in Gieryn and Figert 1986: 80)

of technology or the urgency of political action in particular situations when they are seeking authority that depends on those connections. Yet those same connections are made carefully, to protect the authority of science, and are countered by boundary work aimed at protecting or expanding the authority of engineers and politicians (Jasanoff 1987).

Boundary work is a concept with broad applicability. Norms are not the only resources that can be used to stabilize or destabilize boundaries. Organizations can help to further goals while maintaining the integrity of established boundaries (Guston 1999b; Moore 1996). Boundary work can be routine, occurring when there are no immediate conflicts on the horizon (Kleinman and Kinchy 2003; Mellor 2003). Examples, people, methods, and qualifications are all used in the practical and never-ending work of charting boundaries. Textbooks, courses, and museum exhibits, for example, can establish maps of fields simply through the topics and examples that they represent (Gieryn 1996). In fact, little does not participate in some sort of boundary work, since every particular statement contributes to a picture of the space of allowable statements.

The Place of Norms in Science?

The failings of Merton's functionalist picture of science are instructive. Merton can be seen as asking what science needs to be like, as a social activity, in order for it to best provide certified knowledge. His four norms provide an elegant solution to that problem, and a plausible solution in that they are professed standards of scientific behavior.

Nonetheless, these norms do not seem to describe the behavior of scientists, unless the framework is interpreted very flexibly. But if it is interpreted flexibly then it ceases to do real analytic or explanatory work. Going a little deeper, critics have also challenged the idea that science is a unified institution organized around a single goal or even a set of goals. Instead, the sciences and individual scientific institutions are contested – by governments, corporations, publics, and scientists themselves. Does the idea of an overarching goal, for an entity as large and diffuse as science, even make sense? Could an overarching goal for science have any effect on the actions of individual scientists?

As a result of these arguments, critics suggest that science is better understood as the combined product of scientists acting to pursue their own goals. Merton's norms, then, are ideological resources, available to scientific actors for their own purposes. They serve, combined with formalist epistemologies, as something like an "organizational myth" of science (Fuchs 1993).

Still, we can ask how ideologies like Merton's norms affect science as a whole. It may be that their repeated invocation leads to their having real effects on the shape of scientific behavior: they are used to hold scientists accountable, even if their use is flexible. We might expect, for example, that the repeated demand for universalism will lead to some types of discrimination being unacceptable – shaping the ethics of science. Along with other values, Merton's norms may contribute to what Lorraine Daston (1995) calls the "moral economy" of science. While science as a whole may not have institutional goals, combined actions of individual scientists might shape science to look as though it has goals (Hull 1988). Even though boundaries, in this case boundaries of acceptable behavior, are constructed, they can have real effects.

4

Stratification and Discrimination

An Efficient Meritocracy or an Inefficient Old Boy's Network?

Of the 55 Nobel Laureates working in the United States in 1963, a full 34 of them had studied or collaborated with a total of 46 previous prize-winners (Zuckerman 1977). Not only that, but those who had worked with Nobel Laureates before they themselves had done their important research received their prizes at an average age of 44, compared with an average age of 53 for the others. Clearly scientists tend to form elite groupings. Are these groupings the tip of a merit-based iceberg, or are they artifacts of systems of prestige gone awry? Is the knowledge for which elites are recognized intrinsically and objectively valuable, or does it become so because of its association with elites?

The renaissance philosopher Francis Bacon thought that the inductive method he had set out for science would level the differences among intellectual abilities and allow science to be industrialized. Three hundred years later, Spanish philosopher José Ortega y Gasset claimed that "science has progressed thanks in great part to the work of men astoundingly mediocre, and even less than mediocre." Despite such pronouncements, a small number of authors publish many papers and a great number publish very few. Roughly 10% of all scientific authors produce 50% of all scientific papers (Price 1986). Similarly, a small number of articles are cited many times and a great number are cited very few times: an estimated 80% of citations are to 20% of papers (Cole and Cole 1973). If the number of citations is a measure of influence then these figures suggest that a small number of publications are quite influential, and the vast majority make only a small contribution to future advances. To use citations to represent influence can be misleading (Box 4.1), but, these figures are so striking that even if number of citations

Box 4.1   Do citations tell us about influence?

A number of sociologists have attempted to use citation analysis as a tool to evaluate the importance of particular publications, among other things. But is citation analysis a useful tool for studying scientific communication (Edge 1979)?

An assumption of citation analysis is that citations represent influence. A citation is supposed to be a reference to a publication that provides important background information. This is an idealization. Scientific publications are often misleading about their own histories (e.g. Medawar 1963). An article stemming from an experiment looks as though it recounts the process of experimentation, but is actually a rational reconstruction of that process. It is written to fit into universal categories, so details are cleaned up, idiosyncrasies are left out, and temporal order is changed to create the right narrative. The scientific paper is typically an argument. Its citations, therefore, are vehicles for furthering its argument, not records of influences (Gilbert 1977). Citations are biased toward the types of references that are useful for addressing intended audiences. Meanwhile, informal communication and information that does not need to be cited – or is best not cited – is all left out of publications. Even when citations are not intended to further arguments, they may serve other purposes, perhaps several purposes simultaneously (Case and Higgins 2000). Citation analysis is therefore a poor tool for studying communication and influence (MacRoberts and MacRoberts 1996).

is a crude measure of influence, Bacon and Ortega are wrong about the structure of science.

It is useful to contrast two possible policy implications of such figures, to see some of their practical importance. One possible conclusion, drawn by Jonathan and Stephen Cole, is that scientific funding could be allocated much more efficiently, being placed in the hands of people more likely to write influential, and thus more valuable, articles. Since science is already highly stratified, funding could better reflect that stratification. Most of the 80% less-cited papers need not be written. They may be entirely replaceable, and if so this would reduce the value of the already-small contribution they make: the Coles argue that, given the number of multiple discoveries in the history of science, discoveries happen regularly.

The recommendation to further stratify funding is based on the assumption that citations reflect intrinsic value: if 80% of citations are to 20% of the articles, it is because those 20% are intrinsically more valuable. Stephen Turner and Daryl Chubin (1976) point out that by denying this assumption one can draw very different policy lessons from the same citation data. Perhaps science uses its resources poorly, and the majority of perfectly good articles are ignored. Turner and Chubin's alternative policy recommendations would be aimed at leveling playing fields, reducing the effects of prestige and old-boys networks. The disagreement turns on the question of how accurate science's system of recognition is, a question that is difficult to answer.

One might also raise very different questions about where and how science identifies value. Both the Coles and Turner and Chubin seem to assume that scientific publications are the only significant locus of scientific value. They share a perspective that sees science as essentially a producer of pieces of knowledge. However, it might be argued that scientific skills are of great value – Chandra Mukerji (1989) claims that oceanography gets government support because it produces a pool of skilled oceanographers, who can be called upon to perform particular services. Or, in some cases scientific skills and knowledge contribute directly to technological change, not via published knowledge but via exporting laboratory tools and products to more general use. While neither of these alternatives looks right in general, they show some of the difficulties of a narrow focus on publications.

Contributions to Productivity

No obvious variable has an overwhelming effect on scientists' publication productivity (generally measured in terms of number of publications). However, one survey builds up a picture of the most productive scientist as someone who:

> is driven, has a strong ego, has a history of acting autonomously, plays with ideas, tolerates ambiguity, is at home with abstractions, is detached from non-scientific social relations, reserves mornings for writing, works on routinized problems, is relatively young, went to a prestigious graduate school, had a prestigious mentor, is currently at a prestigious place of employment, has considerable freedom to choose their own problems, has a history of success, and has had previous work cited. (for references see Fox 1983)

A recent study, based on a survey of researchers, has suggested more abstract behavioral categories. The successful researcher is, in order of importance:

Persistent, smart, civil, creative, entrepreneurial, aggressive, tasteful, confident, and adaptable. (Hermanowicz 2006)

Many of the items in these lists are uncontroversial, but for somebody seeking to understand scientific processes they are not particularly helpful. For example, peak productive years for scientists tend to center around their mid-forties, but it is unclear why, or what could be done to increase the productivity of older scientists. Similarly, persistence may indeed be very important, but the sources of persistence are unclear.

Scientists working at prestigious locations have strikingly higher productivity, and institutional prestige is a strong predictor of productivity (e.g. Allison and Long 1990; Long and McGinnis 1981). Do prestigious university departments and independent laboratories select researchers who are intrinsically more productive? Not in general: evidence suggests that differences in productivity emerge *after* people have been hired. Do employers with more prestige select scientists who intrinsically are going to be more productive? Perhaps, though it is difficult to imagine how they would be able to do so as consistently as they do, especially since the academic job market is almost always tight, and the differences in potential between candidates who are hired at more and less prestigious institutions could be expected to be very slight. So, how does prestige contribute to productivity? There are a number of possibilities. Facilities and networking opportunities are typically better at prestigious locations. Prestigious environments might provide more intellectual stimulation than do non-prestigious environments. Prestigious environments might also be ones in which the internal pressure and rewards for productivity are greatest. Or, prestigious environments might provide more motivation by providing more visibility, and therefore more rewards for researchers – if this is right then prestige by itself makes people's research valuable!

The *cumulative advantage* hypothesis ties many of the above variables together. Robert Merton (1973: Ch. 20) coined the term *Matthew Effect* after the line from the Gospel According to St Matthew (13, line 12): "For whosoever hath, to him shall be given, and he shall have more abundance: but whosoever hath not, from him shall be taken away even that he hath." In science, success breeds success. People with some of the identified valuable psychological traits and work habits in the list above are more likely to have a famous mentor, and to study at a prestigious graduate school. Once there, they are more likely to gain employment in a prestigious department. People in better departments may have access to better facilities and intellectual stimulation. Perhaps more importantly, they are more visible and therefore more likely to be cited. Within the social structure of science, citation is a reward: people whose work is cited tend to continue publishing (Lightfield 1971) – in Randall Collins's (1998) terms, citations are sources of emotional energy.

The cumulative advantage hypothesis is attractive not merely because it ties together so many insights. It is also attractive because it straddles a fine line between seeing science's reward structure as merit-based and seeing it as elitist or idiosyncratic. Rewards for small differences, which may or may not be differences in merit, can end up having large effects, including effects on productivity – small differences can become large ones.

Discrimination

Women have made tremendous recent progress in the sciences. However, they are still poorly represented in some fields, and are more poorly represented the more elite the grouping of scientists. In the United States in 2005, women earned 46% of doctoral degrees in the social sciences, natural sciences, and engineering, an increase from 31% from a decade earlier (National Science Board 2008). Yet women earned 49% of doctoral degrees in the biological sciences but only 18% in engineering. Figures for women employed in the sciences reveal, for example, that in the United States in 2006 only 19% of senior faculty in these areas were women, heavily concentrated in the life and social sciences. Such figures represent a marked improvement from 30 years earlier (emphasized, for example, by Long 2001), but they still indicate discrimination within science and engineering (Schiebinger 1999).

Women's participation in science and engineering also varies quite substantially from country to country. Overall, women earn a larger percentage of European doctoral degrees than US ones, but figures range considerably. For example, in 2003 women earned 72% of life science and 45% of physical science doctorates in Italy, but only 47% of life science and 23% of physical science doctorates in Germany (European Commission 2006). Only 15% of senior university faculty in Europe, across all fields, are women. Again, there is geographic variation: 22% of senior medical faculty in the UK are women, versus only 6% in Germany (European Commission 2006). Among countries that are large contributors to research, Japan stands out as particularly gendered, with smaller percentages of women researchers of all kinds than the United States or any European country. And in most countries, women are quite unlikely to be gatekeepers: editors of major journals, chairs of important scientific boards, etc.

| Box 4.2   Women scientists in America |

Margaret Rossiter's (1982, 1995) historical studies of women scientists in the United States present pictures of the struggles of women to work within and change the cultures and institutions of science. For her, an overarching problem that women have faced in science has been that feminization is linked to a lowering of status and masculinization to an increase in status. When, in the 1860s and 1870s, increasing numbers of women attempted to enter scientific fields, men in those fields perceived that their work was losing its masculine status (Rossiter 1982). As a result, they forced women into marginal positions in which they would be "invisible." In the twentieth century women found that they could enter the scientific world much more easily if they entered new fields, such as home economics and nutritional science. Within those fields they could do sociology, economics, biology, and chemistry; however, because of their feminization the fields as wholes were poorly regarded.

Similarly, women's colleges in the United States once provided safe environments for women scientists, in which they could work and establish their credentials. During the expansion and rise of colleges and universities after World War II, women's colleges and less prestigious universities followed the lead of more prestigious universities in hiring more men. The "golden age" of science in the United States after World War II was a "dark age" for women in science (Rossiter 1995).

A qualitative change to women's place in US science happened only in the 1960s. The National Organization for Women and activists like Betty Friedan articulated issues of women's rights in terms that resonated with US ideals. Alice Rossi's 1965 article in *Science*, entitled "Women in Science: Why So Few?" articulated problems women faced in professional life, and made clear that science was no different from other areas. At the same time, there began a legal revolution that resulted in new rights for women, in both education and employment. Resulting anti-discrimination laws and affirmative action programs have been useful resources in women's struggles to "get in" to the worlds of professional science and engineering.

Natalie Angier uses a common metaphor, a pipeline, to describe the problem of women in science and engineering. Women flow through "a pipe with leaks at every joint along its span, a pipe that begins with a high-pressure surge of young women at the source – a roiling Amazon of smart graduate students – and ends at the spigot with a trickle of women prominent enough to be deans or department heads at major universities or to win such honours as membership in the National Academy of Science or even, heaven forfend, the Nobel Prize" (quoted in Etzkowitz et al. 2000: 5). The problem is typically seen as one of leakage: women are leaking out of the pipeline all the way along. However, the pipeline metaphor makes most sense if we pay attention not only to leakage but also to intake, blockage, and filters, or processes of inclusion and exclusion (Etzkowitz et al. 2000; Lagesen 2007). In many scientific fields, there is no roiling Amazon of women graduate students, and those who are there do not merely leak out but face discouragements all along the way. Sara Delamont (1989) describes the problem of equity for women in terms of three problems, "getting in," "staying in," and "getting on" (see also Glover 2000). If science is to move toward equity, women need to enter the pipeline, not leak out, and not hit barriers that prevent their careers from moving forward.

Getting In, Staying In, and Getting On

Science and engineering have masculine images, and the stereotypical scientist is male. These facts shape perceptions of girls' and women's abilities to be scientists (Easlea 1986; Faulkner 2007). Quite young girls may find it difficult to think of themselves as potential scientists, and in many environments they are actively discouraged by teachers, peers, and parents (Orenstein 1994). The result is that by the end of their secondary education, many girls who might have been interested in scientific careers already lack some of the background education they need to study science in colleges and universities. The stereotypes continue. An analysis of advertisements in the journal *Science* shows that "there is more cultural content between the covers of *Science* than we would care to acknowledge" (Barbercheck 2001). Men are more likely to be depicted as scientific heroes than are women; they are also more likely to be depicted as nerds or nonconformists, and more likely to be depicted at play. Words like "simple" and "easy" are more likely to be associated with pictures of women than men, and words like "fast," "reliable," and "accurate" are more likely to be associated with pictures of men than women. And female figures can serve as symbols of nature in a way that male figures cannot.

As graduate students, women may be excluded from key aspects of scientific socialization and training. Etzkowitz, Kemelgor, and Uzzi (2000) refer to the "unofficial Ph.D. program" that runs alongside the official courses, exams, and research. The unofficial PhD can involve everything from study groups, pick-up basketball games, informal mentoring, and simply ongoing conversations and consultation (see also Delamont and Atkinson 2001). Even well-integrated students may be excluded from parts of the unofficial program. One female doctoral student in their study reports: "We would all go to parties together and go and have beer on Friday, but if somebody came in to ask what drying agent to use to clean up THF, they would never ask me. It just wasn't something that would cross their minds" (Etzkowitz et al. 2000: 74). Exclusion from aspects of education can take specific forms in some regions; Indian women doctoral students, for example, may face particular segregation because of cultural traditions (Gupta 2007). Some disciplines have strongly gendered stereotypes of successful PhD students, as well as postdoctoral and more senior researchers (Traweek 1988). Thus, sexism can be part of the everyday texture of scientific life.

Staying in brings its own set of problems. Being hired into good positions may be more difficult for women, simply because they tend not to look like younger versions of the people hiring them, and possibly because of deliberate discrimination. A study of the peer review scores for postdoctoral fellowships by the Swedish Medical Research Council showed that reviewers rated men's accomplishments as overwhelmingly more valuable than apparently equivalent accomplishments by women (Wenneras and Wold 2001). "For a female scientist to be awarded the same competence score as a male colleague, she needed to exceed his scientific productivity by . . . approximately three extra papers in *Nature* or *Science* . . . or 20 extra papers in . . . an excellent specialist journal such as *Atherosclerosis*, *Gut*, *Infection and Immunity*, *Neuroscience*, or *Radiology*" (Wenneras and Wold 2001). Discrimination can take very blatant and extreme forms.

Shirley Tilghman, a successful molecular biologist and President of Princeton University, argues that professional pressures at the beginnings of careers are particularly difficult for women (Tilghman 1998). Research productivity is crucial for senior graduate students and postdoctoral researchers, if they are to land permanent positions. In many systems, new university faculty have time for little except teaching and research until they gain tenure, an all-or nothing hoop through which they must jump after about six years. This pressure usually comes at the time when women are most likely to be interested in having children and spending time with their young families.

Since the stratification of science is shaped by cumulative advantage and disadvantage, early careers are particularly important. The problems faced by women as graduate students and young researchers are particularly important for their long-term success. Young women scientists are often discriminated against at least in small ways (Fox 1991). They tend to have smaller laboratories than male peers hired at the same time, and they tend to have negotiated worse arrangements for start-up grants, for the portion of their research grants that the university takes to cover overhead, for salary, and for teaching duties.

Women are also excluded from informal networks after they have success-fully launched careers. Within their departments they tend not to have informal mentors who can help steer them through difficult processes. Outside their departments they tend not to do as much collaborative research as do men. And they tend not to be as well tied in to informal networks that inform them of unpublished research, leaving them more dependent upon the published literature (Fox 1991). While women have moved into science, they are not *of* the community of science (Cole 1981).

In individual cases such differences can appear trivial, but if competition is strong enough they may have a substantial effect. That is, if the cumulative advantages and disadvantages are strong enough, the small irritants that women face, and may even shrug off, can be enough to keep them from getting on (Cole and Singer 1991).

Finally, there is the question of whether women and men learn and do science differently. On the one hand, there have been a number of claims that there is a "math gene" or some similar biological propensity that accounts for boys' better abilities in mathematics and science (e.g. Benbow and Stanley 1980); as recently as 2005, Harvard University's President Lawrence Summers invoked biology as an explanation for differences in men's and women's success in science and engineering. Given that genes only act in environments, such claims are ideological statements, but in addition they run against recent trends in which boys' and girls' test scores in mathematics are becoming closer and closer (Etzkowitz et al. 2000). On the other hand, a number of feminists and non-feminists have claimed that women and men tend to have distinctive patterns of learning and thinking (e.g. Gilligan 1982; Keller 1985). Women, it is argued, tend to think in more concrete terms than do men, and are thereby better served by problem-centered learning. They also may think more relationally than do men, and therefore produce more complex and holistic pictures of phenomena. This set of issues will be picked up in Chapter 7.

Conclusion

The cumulative advantage and disadvantage hypothesis provides a way of understanding how a nebulous culture can have very concrete effects. For example, science and engineering are and have been dominated by men, and men are still stereotypical scientists and engineers in most fields. Women's participation in science and engineering therefore violates gendered expectations, and women find themselves fighting a number of uphill battles. While no one of these may decisively keep women out of the pipeline, eject women from it, or block their progress through it, their compounded effects may be enough to prevent women from moving smoothly into and through it. The cumulative advantage and disadvantage hypothesis provides a way of understanding how women's stories of succeeding and not succeeding in science can be very different, but nonetheless fit a consistent pattern.

With discrimination in focus, it is difficult to see science as efficiently using its resources. Instead of an efficient meritocracy, science looks more like an inefficient old boys' network, in which people and ideas are deemed important and become important at least partly because of their place in various social networks. In the broadest sense, then, elite groups are socially constructed, rather than being mere reflections of natural and objective hierarchies. Undoubtedly elites produce more or better knowledge than do non-elites: they have cumulative advantages, and are more likely to be positively reinforced for their work. However, the fact of discrimination raises the question of whether some of the knowledge for which elites are recognized becomes important because of its association with them, and not vice versa. It also raises the question of whether science could be significantly different were it more egalitarian.

Box 4.3   Women at the margins of science

The places of women just outside the formal borders of science and technology are interesting for what they show about women's informal contributions. The tradition of studying under-recognized women in science and technology is well established (e.g. Davis 1995; Schiebinger 1999), but it has also taken some novel turns aimed more at understanding cultures surrounding science and technology than simply in recovering contributions. For example, Jane Dee, the wife of Elizabethan alchemist and mathematician John Dee, not only had to manage an "experimental household," but she had to manage her own location between Elizabethan society and natural philosophy. It was a location that was difficult to manage, because natural philosophy was done in the home, where Jane Dee had to act as a full partner to her husband, yet Elizabethan natural philosophy was chaotic, and its place not well defined, creating a great potential for disorder in the household (Harkness 1997).

Looking at a more public space, middle- and upper-class women made up a large part of the audience at meetings of the British Association for the Advancement of Science (BAAS) in the nineteenth century, and they were accepted as such – more than working-class men, for whom the BAAS organized special lectures. However, even when the occasional woman started participating, women as a group remained firmly in the role of audience, and even tended to see themselves as passive observers (Higgitt and Withers 2008).

The nineteenth century saw liminal spaces open up for women to participate in science in women's colleges. For example, women's colleges in the United States developed a distinctive subculture of physiological teaching and research (Appel 1994). Although that was primarily aimed at practical applications, it allowed some women students to develop scientific knowledge and experimental skills that could then gain them access to graduate education elsewhere. To take another example, the Balfour Biological Laboratory for Women at Cambridge University served women students at a time when they were being excluded from lectures and laboratories elsewhere on campus (Richmond 1997). Although it did not create high-prestige opportunities for women, the Laboratory was a small-scale natural science environment parallel to those of the men. It had the resources for women students to learn the skills needed for demanding natural science exams, and was a space in which there were even opportunities for women laboratory demonstrators and lecturers.

5

The Strong Programme and the Sociology of Knowledge

The Strong Programme

In the 1970s a group of philosophers, sociologists, and historians based in Edinburgh set out to understand not just the organization but the *content* of scientific knowledge in sociological terms, developing the "strong programme in the sociology of knowledge" (Bloor 1991 [1976]; Barnes and Bloor 1982; MacKenzie 1981; Shapin 1975). The most concise and best-known statement of the programme is David Bloor's "four tenets" for the sociology of scientific knowledge:

1. It would be causal, that is, concerned with the conditions which bring about belief or states of knowledge. . . .
2. It would be impartial with respect to truth and falsity, rationality or irrationality, success or failure. Both sides of these dichotomies will require explanation.
3. It would be symmetrical in its style of explanation. The same types of cause would explain, say, true and false beliefs.
4. It would be reflexive. In principle its patterns of explanation would have to be applicable to sociology itself. (Bloor 1991 [1976], 5)

These represent a bold but carefully crafted statement of a naturalistic, perhaps scientific, attitude toward science and scientific knowledge, which can be extended to technological knowledge as well. Beliefs are treated as objects, and come about for reasons or causes. It is the job of the sociologist of knowledge to understand these reasons or causes. Seen as objects, there is no *a priori* distinction between beliefs that we judge true and those false, or those rational and irrational; in fact, rationality and irrationality are themselves objects of study. And there is no reason to exempt sociology of knowledge itself from sociological study.

Since the strong programme STS has been concerned with showing how much of science and technology can be accounted for by the *work* done by scientists, engineers, and others. To do so, the field has emphasized Bloor's stricture of symmetry, that beliefs judged true and false or rational and irrational should be explained using the same types of resources. Methodological symmetry is a reaction against an asymmetrical pattern of explanation, in which true beliefs require internal and rationalist explanations, whereas false beliefs require external or social explanations. Methodological symmetry thus opposes a variety of Whig histories of science (Chapter 2), histories resting on the assumption that there is a relatively unproblematic rational route from the material world to correct beliefs about it. Whig history assumes a foundationalism on which accepted facts and theories ultimately rest on a solid foundation in nature (Box 2.2). As the problems of induction described so far show, there is no guaranteed path from the material world to scientific truths, and no method identifies truths with certainty, so the assumption is highly problematic. Truth and rationality should not be privileged in explanations of particular pieces of scientific knowledge: the same types of factors are at play in the production of truth as in the production of falsity. Since ideology, idiosyncrasy, political pressure, etc., are routinely invoked to explain beliefs thought false, they should also be invoked to explain beliefs thought true.

Although there are multiple possible interpretations of methodological symmetry, in practice it is often equivalent to agnosticism about scientific truths: we should assume that debates are open when we attempt to explain closure. The advantage of such agnosticism stems from its pushes for ever more complete explanations. The less taken for granted, the wider the net cast to give a satisfactory account. In particular, too much rationalism tends to make the analyst stop too soon. Many factors cause debates to close, for science to arrive at knowledge or technology to stabilize, and therefore multiple analytic frameworks are valuable for studying science and technology.

In addition to statements of the potential for sociology of knowledge, strong programmers Bloor (1991) and Barry Barnes (1982; also Barnes, Bloor, and Henry 1996) have offered a new and useful restatement of the problem of induction. The concept of "finitism" is the idea that each application of a term, classification, or rule requires judgments of similarity and difference. No case is or is not the same as cases that came before it in the absence of a human decision about sameness, though people observe and make decisions on the basis of similarities and differences. Terms, classifications, and rules are extended to new cases, but do not simply apply to new cases before their extension.

Actors and observers normally do not feel or see the open-endedness that finitism creates. According to Barnes and Bloor, this is because different kinds of social connections fill most of the gaps between past practices and their extension to new cases. That is, since there is no logic that dictates how a term, classification, or rule applies to a new case, social forces push interpretations in one direction or another. This "sociological finitism" (see Box 5.1) opens up a large space for the sociology of knowledge! As Bloor says:

Can the sociology of knowledge investigate and explain the very content and nature of scientific knowledge? Many sociologists believe that it cannot . . . They voluntarily limit the scope of their own enquiries. I shall argue that this is a betrayal of their disciplinary standpoint. All knowledge, whether it be in the empirical sciences or even in mathematics, should be treated, through and through, as material for investigation. (Bloor 1991 [1976]: 1)

Box 5.1   Sociological finitism

The argument for finitism is a restatement of Wittgenstein's argument about rules (Box 3.2). Rules are *extended* to new cases, where extension is a process. Rules therefore change meaning as they are applied. Classifications and the applications of terms are just special cases of rules.

The application of finitism can be illustrated through an elegant case in the history of mathematics analyzed by Imre Lakatos (1976), and re-analyzed by Bloor (1978). The mathematics is straightforward and helpfully presented, especially by Lakatos. The case concerns a conjecture due to the mathematician Leonard Euler: for polyhedra, $V - E + F = 2$, where F is the number of faces, E the number of edges, and V the number of vertices. Euler's conjecture was elegantly and simply proven in the early nineteenth century, and on the normal image of mathematics that should have been end of the story. However, quite the opposite occurred. The proof seemed to prompt counter-examples, cases of polyhedra for which the original theorem did not apply!

Some mathematicians took the counter-examples as an indictment of the original conjecture; the task of mathematics was then to find a more complicated relationship between V, E, and F that preserved the original insight, but was true for all polyhedra. Other mathematicians took the counter-examples to show an unacceptable looseness of the category *polyhedra*; the task of mathematics was then to find a definition of *polyhedra* that made Euler's conjecture and its proof correct, and that ruled the strange counter-examples as "monsters." Still others saw opportunities for interesting classificatory work that preserved the original conjecture and proof while recognizing the interest in the counter-examples.

It seems reasonable to say that at the time there was no correct answer to the question of whether the counter-examples were polyhedra. Despite the fact that mathematicians had been working with polyhedra for millennia, any of the responses could have become correct, because the meaning of polyhedra had to change in response to the proof and counter-examples. On Bloor's analysis, the types of societies and institutions in which mathematicians worked shaped their responses, determining whether they saw the strange counter-examples as welcome new mathematical objects with just as much status as the old ones, as mathematical pollution, or simply as new mathematical objects to be integrated into complex hierarchies and orders.

Interest Explanations

The four tenets of the strong programme do not set limits on the resources available for explaining scientific and technological knowledge, and do not establish any preferred styles of explanation. In particular, they do not distinguish between externalist explanations focused on social forces and ideologies that extend beyond scientific and technical communities, on the one hand, and internalist explanations focused on forces that are endemic to those communities, on the other. This distinction is not perfectly sharp or invariant, nor are many empirical studies confined to one or other side of the divide.

Though the strong programme can cover both externalist and internalist studies, it was early on strongly associated with the former. When the strong programme was articulated in the 1970s, historians of science were having historiographical debates about internal and external histories, particularly in the context of Marxist social theory. The strong programme slid too neatly into the existing discussion.

Externalist historians or sociologists of science attempt to correlate and connect broad social structures and events and more narrow intellectual ones. Some difficulties in this task can be seen in an exchange between Steven Shapin (Shapin 1975) and Geoffrey Cantor (Cantor 1975). Shapin argues that the growth of interest in phrenology – psychology based on external features of people's heads – in Edinburgh in the 1820s was related to a heightened class struggle there. The Edinburgh Phrenological Society and its audiences for lectures on phrenology were dominated by members of the lower and middle classes. Meanwhile, the Royal Society of Edinburgh, which had as members many of the strongest critics of phrenology, was dominated by the upper classes. Reasons for these correlations are somewhat opaque, but Shapin claims that they are related to connections between phrenology and reform movements.

Cantor makes a number of criticisms of Shapin's study: (1) Class membership is not a clear-cut matter, and so the membership in the societies in question may not be easily identified as being along class lines. In addition, on some interpretations there was considerable *overlap* in membership of the two societies, raising the question of the extent to which the Phrenological Society could be considered an outsider's organization. (2) Shapin does not define 'conflict' precisely, and so does not demonstrate that there was significantly more conflict between classes in Edinburgh in the 1820s than there had been at some other time. (3) While the overall picture of membership of the two societies may look different, they had similar percentages of members coming from some professional groups. To make this point vivid, Cantor calls for a social explanation of the *similarity* of composition of the two societies. A correlation may not be evidence for anything.

While Cantor's criticisms of Shapin are specific to this particular account, they can be applied to other interest-based accounts. Many historical studies follow the same pattern: They identify a scientific controversy in which the debaters on each side can be identified. They identify a social conflict, the sides of which can be correlated to the sides of the debate. And finally, they offer an explanation to connect the themes of the scientific debate and those of the social conflict (e.g. MacKenzie 1978; Jacob 1976; Rudwick 1974; Farley and Geison 1974; Shapin 1981; Harwood 1976, 1977).

Exactly the same problems may face many internalist accounts, but are not so apparent because the posited social divisions and conflicts often seem natural to science and technology, and so are more immediately convincing as causes of beliefs. Conflicts between physicists with different investments in mathematical skills (Pickering 1984), between natural philosophers with different models of scientific demonstration (Shapin and Schaffer 1985), or between proponents of different methods of making steel (Misa 1992) involve more immediate links between interests and beliefs, because the interests are apparently internal to science and technology.

Steve Woolgar (1981) has developed a further criticism of interest-based explanations. Analysts invoke interests to explain actions even when they cannot display a clear causal path from interests to actions. To be persuasive, then, analysts have to isolate a particular set of interests as dominant, and independent of the story being told. However, there are indefinitely many potential interests capable of explaining an action, so any choice is under-determined. Woolgar is criticizing "social realism," the assumption that aspects of the social world are determinate (even if aspects of the natural world are not). Woolgar, advancing the reflexive part of the strong programme, points out that accounts in STS rhetorically construct aspects of the social world, in this case interests, in exactly analogous ways as scientists construct aspects of the natural world. STS should make social reality and natural reality symmetrical, or should justify their lack of symmetry. Actor-network theory attempts the former (see Chapter 8; Latour 1987; Callon 1986). Methodo-logical relativism adopts the latter strategy (Collins and Yearley 1992).

Interest-based patterns of analysis thus face a number of problems: (1) analysts tend to view the participants in the controversy as two-dimensional characters, having only one type of social interest, and a fairly simple line of scientific thought; (2) they tend to make use of a simplified social theory, isolating few conflicts and often simplifying them; (3) it is difficult to show causal links between membership in a social group and belief; and (4) interests are usually taken as fixed, and society as stable, even though these are as constructed and flexible as are the scientific results to be explained.

Despite these problems, STS has not abandoned interest-based explana-tions. They are too valuable to be simply brushed aside. First, interest explanations are closely related to rational choice explanations, in which actors try to meet their goals. Rational choices need to be situated in a context in which certain goals are highlighted, and the choices available to reach those goals are narrowed. The difficult theoretical problems are answered in practical terms, by more detailed and cautious empirical work. Second, researchers in STS have paid increasing attention to scientific and technical cultures, especially material cultures, and how those cultures shape options and choices. They have emphasized clear internal interests, such as interests in particular approaches or theories. As a result, researchers in STS have shown how social, cultural, and intellectual matters are not distinct. Instead, intellectual issues have social and cultural ones woven into their very fiber. Within recognized knowledge-creating and knowledge-consuming cultures and societies this is importantly the case (see Box 5.3), but it is also the case elsewhere. Third, while situating these choices involves rhetorical work on the part of the analyst, this is just the sort of rhetorical work that any explanation requires. Woolgar's critique is not so much of interests, but is a commentary on explanation more broadly (Ylikoski 2001).

Box 5.2   Testing technologies

A test of a technology shows its capabilities only to the extent that the circumstances of the test are the same as "real-world" circumstances. According to the finitist argument, though, this issue is always open to interpretation (Pinch 1993a; Downer 2007). We might ask, with Donald MacKenzie (1989), how accurate are ballistic missiles? This is an issue of some importance, not least to the militaries and governments that control the missiles. As a result, there have been numerous tests of unarmed ballistic missiles. Although the results of these tests are mostly classified, some of the debates around them are not.

As MacKenzie documents, critics of ballistic missile tests point to a number of differences between test circumstances and the presumed real circumstances in a nuclear war. For example, in the United States, at least until the 1980s when MacKenzie was doing his research, most inter-continental ballistic missiles had been launched from Vandenberg Air Force Base in California to the Kwajalein Atoll in the Marshall Islands. The variety of real launch sites, targets, and even weather was missing from the tests. In a test, an active missile is selected at random from across the United States, sent to Texas where its nuclear warhead is removed and replaced by monitoring equipment, and where the missile is wired to be blown up in case of malfunction, and finally sent on to California. To make up for problems introduced in transportation, there is special maintenance of the guidance system prior to the launch. It is then launched from a silo that has itself been well maintained and well studied. Clearly there is much for critics of the tests to latch on to. A retired general says that "About the only thing that's the same [between the tested missile and the Minuteman in the silo] is the tail number; and that's been polished" (MacKenzie 1989: 424). Are the tests representative, then? Unsurprisingly, interpretations vary, and they are roughly predictable from the interests and positions of the interpreters.

Trevor Pinch (1993a) gives a general and theoretical treatment of MacKenzie's insights on technological testing. He points out that whether tests are prospective (testing of new technologies), current (evaluating the capabilities of technologies in use), or retrospective (evaluating the capabilities of technologies typically after a major failure), in all three of these situations projections have to be made from test circumstances onto real circumstances. Judgments of similarity have to be made, which are potentially defeasible. It is only via human judgment that we can project what technologies are capable of, whether in the future, in the present, or the past.

Knowledge, Practices, Cultures

From the perspective of many of its critics, the strong programme rejects truth, rationality, and the reality of the material world (e.g. Brown 2001). It is difficult to make sense of such criticisms. The strong programme does not reject any of these touchstones, but rather shows how by themselves truth, rationality, and the material world have limited value in explaining why one scientific claim is believed over another. In order to understand scientific belief, we need to look to interpretations and the rhetorical work to make those interpretations stick (Barnes et al. 1996).

For other critics, the strong programme retains too much commitment to truth and the material world. Strong programmers (e.g. Barnes et al. 1996) reject what are sometimes seen as idealist tendencies in STS. It may be that arguments on this issue weaken STS's commitment to methodological agnosticism about scientific truths – probably the strong programme's largest contribution to the field. Whether this is right remains an open question.

Finally, the strong programme has been criticized for being too committed to the reality and hardness of the social world: it is seen as adopting a foundationalism in the social world to replace the foundationalism in the material world that it rejects. As Woolgar argues, there is no reason to see interests as less malleable than anything to be explained. The critique of interests has been amplified by arguments that interests are translated and modified as scientific knowledge and technological artifacts are made (Latour 1987; Pickering 1995). Society, science, and technology are produced together, and by the same processes – this results in a "supersymmetry" (Callon and Latour 1992).

Since the 1970s, the pendulum in STS, and elsewhere in the humanities and social sciences, has swung from emphasizing structures to emphasizing agency (e.g. Knorr Cetina and Cicourel 1981). The strong programme was often associated with structuralist positions, though its statements leave this issue entirely open. Thus as a philosophical underpinning for STS the strong programme has been supplemented by others: constructivism (e.g. Knorr Cetina 1981), the empirical programme of relativism (Collins 1991 [1985]), actor-network theory (e.g. Latour 1987), symbolic interactionism (e.g. Clarke 1990), and ethnomethodology (e.g. Lynch 1985). These positions are briefly described in other chapters. For the most part, this book avoids anatomizing theoretical positions and disputes, except as they directly inform its main topics, but readers interested in philosophical and methodological underpinnings might look at these works and many others to see some of the contentious debates around foundational issues.

Box 5.3   Pierre Bourdieu and an agonistic science

The French anthropologist Pierre Bourdieu (1999 [1973]) argued that we can understand scientific achievements as resulting from the interplay of researchers on a scientific field. Bourdieu introduced the term *cultural capital*, as a counterpart to *social capital* (a term that his use helped to make common) and economic capital. Actors come to a field with a certain amount of each of these forms of capital, and develop and deploy them to change their relative status within the field. This is true even of scientific fields: Bourdieu says, "The 'pure' universe of even the 'purest' science is a social field like any other, with its distribution of power and its monopolies, its struggles and strategies, interests and profits." Action in a field is agonistic.

All scientific moves of an actor are simultaneously moves on the field: "Because all scientific practices are directed toward the acquisition of scientific authority . . . , what is generally called 'interest' in a particular scientific activity . . . is always two-sided" (Bourdieu 1999). Every idea is also a move to increase capital; it is artificial to separate the pursuit of ideas from the social world.

There are, then, many different strategies for increasing capital. For example: "Depending on the position they occupy in the structure of the field . . . the 'new entrants' may find themselves oriented either towards the risk-free investments of succession strategies . . . or towards subversion strategies, infinitely more costly and more hazardous investments which will not bring them the profits . . . unless they can achieve a complete redefinition of the principles legitimating domination."

Bourdieu's work is isolated from other work in STS, and has been picked up by relatively few researchers. However, we can see his agonistic theory of science as respecting the principles of the strong programme in an individualistic way, explaining the content of science as the result of actions of individual researchers.

The strong programme provided an argument that one can study the content of science and technology in social and cultural terms, providing an initial justification for STS. As the field has developed, scientific and technological practices themselves have become interesting, not just as steps to understanding knowledge. Similarly, science as an activity has become an important and productive locus of study. Key epistemic concepts – such as experimentation, explanation, proof, and objectivity – understood in terms of the roles they play in scientific practice, have become particularly interesting. The strong programme's attention to explaining pieces of knowledge in social terms now seems a partial perspective on the project of understanding science and technology, even if it is a crucial foundation.

6

The Social Construction of Scientific and Technical Realities

What Does "Social Construction" Mean?

The term *social construction* started to become common in STS in the late 1970s (e.g. Mendelsohn 1977; van den Daele 1977; Latour and Woolgar 1979). Since then, important texts have claimed to show the social construction (or simply the construction) of facts, knowledge, theories, phenomena, science, technologies, and societies. Social constructivism, then, has been a convenient label for what holds together a number of different parts of STS. And social constructivism has been the target of fierce arguments by historians, philosophers, and sociologists, usually under the banner of *realism*.

For STS, social constructivism provides three important assumptions, or perhaps reminders. First is the reminder that science and technology are importantly *social*. Second is the reminder that they are *active* – the construction metaphor suggests activity. And third is the reminder that science and technology do not provide a direct route from nature to ideas about nature, that the products of science and technology are *not themselves natural*. While these reminders have considerable force, they do not come with a single interpretation. As a result, there are many different "social constructions" in STS, with different implications. This chapter offers some categories for thinking about these positions, and the realist positions that stand in opposition to them.

The classification here is certainly not the only possible one. For example, one of Ian Hacking's goals in his book on constructivism (Hacking 1999) is to chart out the features of a strong form of social constructivism, to provide the sharpest contrast to realism. The features he identifies are: the contingency of facts, nominalism (discussed below) about kinds, and external explanations for stability. On his analysis, a social constructivist account of, say, an established scientific theory will tend toward the position that: the theory was not the only one that could have become established, the categories used in the theory are human impositions rather than natural kinds, and the reasons for the success of the theory are not evidential reasons.

The goals of this chapter are somewhat different. Rather than providing a single analysis or a sharp contrast to realism, the chapter pulls social constructivism apart into many different types of claims. Some of these are clearly anti-realist and some are not. While there are some general affinities among the claims, they also have unique features.

Although most of the forms of constructivism described below are opposed to forms of realism, there is less need for an analogous list of realisms. This does not mean that realism is any more straightforward. Realism typically amounts to an intuition that truths are more dependent upon the natural world than upon the people who articulate them: there is a way that the world is, and it is possible to discover and represent it reasonably accurately. Realists disagree, though, over what science realistically represents, and over what it means to realistically represent something. They disagree about whether the issue is fundamentally one about knowledge or about things. And there is no good account of the nature of reality, the conditions that make real things real – for this reason, realism is probably less often a positive position than a negative one, articulated in opposition to one or another form of anti-realism.

1   The social construction of social reality

Ideas of social construction have many origins in classic sociology and philosophy, from analyses by Karl Marx, Max Weber, and Émile Durkheim, among others. STS imported the phrase "social construction" from Peter Berger and Thomas Luckmann's *The Social Construction of Reality* (1966), an essay on the sociology of knowledge. That work provides a succinct argument for why the sociology of knowledge studies the social construction of reality:

> Insofar as all human "knowledge" is developed, transmitted and maintained in social situations, the sociology of knowledge must seek to understand the processes by which this is done in such a way that a taken-for-granted "reality" congeals for the man in the street. In other words, we contend that *the sociology of knowledge is concerned with the analysis of the social construction of reality.* (Berger and Luckmann 1966: 3)

The subject that most interests Berger and Luckmann, though, is social reality, the institutions and structures that come to exist because of people's actions and attitudes. These features of the social world exist because significant numbers of people act as if they do. Rules of polite behavior, for example, have real effects because people act on them and act with respect to them. "Knowledge about society is thus a realization in the double sense of the word, in the sense of apprehending the objectivated social reality, and in the sense of ongoingly producing this reality" (Berger and Luckmann 1966: 62). Yet features of the social world become independent because we cannot "wish them away" (1966: 1).

The central point here, or at least the central insight, is one about the metaphysics of the social world. To construct an X in the social world we need only that: (a) knowledge of X encourages behaviors that increase (or reduce) other people's tendency to act as though X does (or does not) exist; (b) there is reasonably common knowledge of X; and (c) there is transmission of knowledge of X. Given these conditions, X cannot be "wished away," and so it exists. For example, gender is real, because it is difficult not to take account of it. Gender structures create constraints and resources with which people have to reckon. As a result, treating people as gendered tends to create gendered people. Genders have causal powers, which is probably the best sign of reality that we have. At the same time, they are undoubtedly not simply given by nature, as historical research and divergences between contemporary cultures show us.

Box 6.1 The social construction of the discovery of the laws of genetics

Discoveries are important to the social structure of science, because more than anything else, recognition is given to researchers for what they discover. This motivates priority disputes, which are sometimes fierce (Merton 1973). Yet it is often difficult to pinpoint the moment of discovery. Was oxygen discovered when Joseph Priestley created a relatively pure sample of it, when Antoine Lavoisier followed up on Priestley's work with the account of oxygen as an element, or at some later point when an account of oxygen was given that more or less agrees with our own (Kuhn 1970)? Such questions lead Augustine Brannigan (1981) to an *attributional* model of discoveries: discoveries are not events by themselves, but are rather events retrospectively recognized as origins (see Prasad 2007 for a similar model of inventions).

The history of Mendelian genetics nicely illustrates the attributional model. On the standard story, Gregor Mendel was an isolated monk who performed ingenious and simple experiments on peas, to learn that heredity is governed by paired genes that are passed on independently, and can be dominant or recessive. His 1866 paper was published in an obscure journal and was not read by anybody who recognized its importance until 1900, when Hugo de Vries, Carl Correns, and Erich Tschermak came upon it in the course of their similar studies. Against this story, Brannigan shows that Mendel's paper was reasonably well cited before 1900, but in the context of agricultural hybridization. In that context, its main result, the 3:1 ratio of characteristics in hybrids, was already known. Mendel was read as replicating that result, and offering a formal explanation of it. Indeed, judging by his presentation of them, Mendel did not seem to recognize his results or theory as a momentous discovery.

Though de Vries had read Mendel's paper before 1900, the first of his three publications on his own experiments and the new insights on genetics makes no mention of it. Correns, pursuing a similar line of research, read de Vries's first publication, and quickly wrote up his own, labeling the discovery "Mendel's Law." Recognizing that he had lost the race for priority, Correns assigned it to an earlier generation – though had Correns read papers from one or two generations earlier still, the title could have gone to any of a number of potential discoverers. De Vries's second and third publications accept the priority of Mendel in awkward apparent afterthoughts, but grumble about the obscurity of Mendel's paper.

Mendel, then, became the discoverer of his laws of genetics as a result of a priority dispute. His work on hybridization was pulled out of that context and made to speak to the newly important questions of evolution. Mendel discovered his eponymous laws in 1866, but only as a result of the events of 1900.

For STS, knowledge, methods, epistemologies, disciplinary boundaries, and styles of work are all key features of scientists' and engineers' social landscapes. To say that these objects are socially constructed in this sense is simply to say that they are *real* social objects, though contingently real. Ludwik Fleck, a key forerunner of STS, already makes these points in his *Genesis and Development of a Scientific Fact* (1979 [1935]). A scientist or engineer who fails to account for a taken-for-granted fact in his or her studies may encounter resistance from colleagues, which shows the social reality of that fact. To point this out is in no way to criticize science and technology. The social reality of knowledge and the practices around knowledge is a precondition of progress. If nothing is reasonably solid, then there is nothing on which to build.

STS has tended to add an active dimension to this metaphor, studying processes of social construction. Claims do not just spring from the subject matter into acceptance, via passive scientists, reviewers, and editors. Rather, it takes work for them to become important. For example, Latour and Woolgar (1979) chronicle the path of the statement "TRF (Thyrotropin Releasing Factor) is Pyro-Glu-His-Pro-NH$_2$," as it moves from being *near nonsense* to *possible* to *false* to *possibly true* to a *solid fact*. Along the way, they chart out the different operations that can be done on a scientific paper, from ignoring it, citing it positively, citing it negatively, questioning it (in stronger and weaker ways), and ignoring it because it is universally accepted. Scientists, and not just science, construct facts.

2   The construction of things and phenomena

Not only representations and social realities are constructed. Perhaps the most novel of STS's constructivist insights is that many of the things that scientists and engineers study and work with are non-natural. The insight is not new – it can be found in Gaston Bachelard's (1984 [1934]) concept of *phenomenotechnique,* and even in the work of Francis Bacon written in the 1600s – but it is put forcefully by researchers writing in the 1980s.

Karin Knorr Cetina writes that "nature is not to be found in the laboratory" (Knorr Cetina 1981). "To the observer from the outside world, the laboratory displays itself as a site of action from which 'nature' is as much as possible excluded rather than included" (Knorr Cetina 1983). For the most part, the materials used in scientific laboratories are already partly prepared for that use, before they are subjected to laboratory manipulations. Substances are purified, and objects are standardized and even enhanced. Chemical laboratories buy pure reagents, geneticists might use established libraries of DNA, and engineered animal models can be invaluable.

Once these objects are in the laboratory, they are manipulated. They are placed in artificial situations, to see how they react. They are subjected to "trials of strength" (Latour 1987) in order to characterize their properties. In the most desirable of situations scientists create phenomena, new stable objects of study that are particularly interesting and valuable (Bogen and Woodward 1988) – "most of the phenomena of modern physics are manufactured" (Hacking 1983).

The result of these various manipulations is that knowledge derived from laboratories is knowledge about things that are distinctly non-natural. These things are constructed, by hands-on and fully material work. We will return to this form of construction in Chapter 14.

In terms of technology, there is nothing the least striking about this observation. Whereas sciences are presumed to display the forms of nature exactly as they are, technology gives new shape and form to old materials, making objects that are useful and beautiful. The fact that technology involves material forms of construction, leaving nature behind, is entirely expected.

3   The scientific and technological construction of material and social environments

Scientific facts and technological artifacts can have substantial impacts on the material and social world – that is the source of much of the interest in them. As such, we can say that science and technology contribute to the construction of many environments.

The effects of technology can be enormous, and can be both intended and unintended. The success of gasoline-powered automobiles helped to create suburbs and the suburban lifestyle, and to the extent that manufacturers have tried to increase the suburban market, these are intentional effects. Similarly, the shape of computers, computer programs and networks are created with their social effects very much in mind: facilitating work in dispersed environments, long-distance control, or straightforward military power (Edwards 1997).

Science, too, shapes the world. Research into the causes of gender differences, for example, has the effect of naturalizing those differences. And there is tremendous public interest in this area, so biological research on genes linked to gender, on the gender effects of hormones, and on brain differentiation between men and women tends to be well reported (e.g. Nelkin and Lindee 1995). More often than not, it is reported to emphasize the inevitability of stereotypical gendered behavior. There is good reason to expect that this reporting has effects on gender itself.

Science also shapes policy. Government actions are increasingly held accountable to scientific evidence. Almost no action, whether it is in areas of health, economy, environment, or defense, can be undertaken unless it can be claimed to be supported by a study. Scientific studies, then, have at least some effect on public policies, which have at least some effect on the shapes of the material and social world. Science, as well as technology, then, contributes to the construction of our environments.

4   The construction of theories

The most straightforward use of the social construction metaphor in STS describes scientists and engineers constructing accounts, models, and theories, on a basis of data, and methods for transforming data into representations. Science is constructive in a geometrical sense, making patterns appear given the fixed points that practice produces.

Box 6.2   Realism and empiricism

Whether one should believe scientific theories, or merely see them as good working tools, has been one of the most prominent questions in the philosophy of science since the early twentieth century. Most philosophers agree that science's best theories are impressive in their accuracy, but they disagree about whether that empirical success is grounds for believing the theories are true.

The classic *empiricist* argument against truth starts from the claim that all of the evidence for a scientific theory is from empirical data. Therefore, given two theories that make the same predictions, there can be no empirical evidence to tell the difference between the two. But any theory is only one of an infinite number of empirically equivalent theories, so there is no reason to think that it is true. Truth in the standard sense is superfluous (e.g. van Fraassen 1980; Misak 1995).

Scientific realists can challenge empiricists' starting assumption, and argue that data is not the only evidence one can have for a theory. For example, it is desirable that theories be consistent with metaphysical commitments. As a result, realists can challenge the assumption that there are typically an infinite number of equivalent theories, because only a few theories are plausible. And they can argue that there is no way to make sense of the successes of science without reference to the truth or approximate truth of the best theories (see, e.g., Leplin 1984; Papineau 1996). One of the best-developed versions of the latter argument is due to Richard Boyd (e.g. 1985). Boyd argues that we have good reason to believe "what is implicated in instrumentally reliable methodology" (Boyd 1990: 186). The truth of background theories is the best explanation of the success of scientific methods. The strategy is to focus not on how successful theories are at making predictions or accounting for data, but on how successful background theories are in shaping research, which then produces reliable theories.

That science constructs representations on top of data is roughly the central claim made by logical positivism (Chapter 1). For positivism, there is an essential contingency to scientific theories and the like. For any good scientific theory, one can create others that do equally good jobs of accounting for the data. Therefore, contrary to realist accounts (Box 6.2) we should not believe theoretical accounts, if to believe them means committing ourselves to their truth. Bas van Fraassen, whose work is positivist in spirit, says of the metaphor, "I use the adjective 'constructive' to indicate my view that scientific activity is one of construction of rather than discovery: construction of models that must be adequate to the phenomena, and not discovery of truth concerning the unobservable" (van Fraassen 1980: 5). However, for positivists, the contingency of scientific representations is largely eliminated by prior decisions about frameworks; *logical* positivism adopted the assumption that scientists first make decisions about the logical frameworks within which they work, before they operate within those frameworks.

Given the problems of induction we have seen, this process cannot be a purely methodical or mechanical one. There is no one way to develop good theoretical pictures on a basis of finite amounts of data, no direct route from nature to accounts of nature. When Knorr Cetina, under the rubric of "constructivism," explains why particular theories are successful, she looks to established practices, earlier decisions, extensions of concepts, tinkering, and local contingencies (Knorr Cetina 1977, 1979, 1981). Representations of nature are connected to nature, but do not necessarily correspond to it in any strong sense.

Controversies show the value of bottom-up accounts of contingency (Chapter 11). By definition, scientific and technical controversies display alternative representations, alternative attempts to construct theories and the like. They can also display some of the forces that lead to their closure. For example, a choice between Newton's and Leibniz's metaphysics may have been related to political circumstances (Shapin 1981). Or, the particular resolution of the dispute between Louis Pasteur and Félix Pouchet over spontaneous generation was shaped by the composition of the prize committee that decided it (Farley and Geison 1974). Scientific and technological theories, then, are constructed with reference to data, but are not implied by that data.

5   Heterogeneous construction

Successful technological work draws on multiple types of resources, and simultaneously addresses multiple domains, a point that will be developed in Chapters 8 and 9. The entrepreneurial engineer faces technical, social, and economic problems all at once, and has to bind solutions to these problems together in a configuration that works. In helping to develop an artifact, then, the engineer is helping to produce knowledge, social realities, and material and social things. While the various constructions can be parceled out analytically, in practice they are bound together.

Builders of technology do *heterogeneous engineering* (Law 1987). They have to simultaneously build artifacts and build environments in which those artifacts can function – and, typically, neither of these activities can be done on their own. Technologists need to combine raw materials, skills, knowledge, and capital, and to do this they must enroll any number of actors, not all of whom may be immediately compatible. Technologists have the task of building stable networks involving diverse components.

Scientific work is also heterogeneous. Actor-network theory (Chapter 8) is a theory of "technoscience," in which scientists and engineers are separated only by disciplinary boundaries. Like engineers, scientists construct networks, the larger and more stable the better. They both construct order, because stable networks create an orderly world. These networks are heterogeneous in the sense that they combine isolated parts of the material and social worlds: laboratory equipment, established knowledge, patrons, money, institutions, etc. Together, these create the successes of technoscience, and so no one piece of a network can determine the shape of the whole.

What we might call *heterogeneous construction* is the simultaneous shaping of the material and social world, to make them fit each other, a process of "co-construction" (Taylor 1995). Heterogeneous construction can involve all of the other types of construction mentioned to this point,

---

Box 6.3   The heterogeneous construction of the Pap smear

Monica Casper and Adele Clarke (1998) show how the Pap smear became the "right tool for the job" of screening for cervical cancer through a process that we could see as heterogeneous construction. It has its origins in George Papanicolaou's study, published in 1917, on vaginal smears as indicators of stages in the estrous cycle of guinea pigs. Over the following decade, Papanicolaou investigated other uses of the smears, eventually discovering that he could detect free-floating human cancer cells. His presentation of that finding, in 1928, was met with little enthusiasm: the results were not convincing, pathologists were not used to looking at free-floating cells, and gynecologists were uninterested in cancer. Papanicolaou abandoned the smear as a cancer test for another decade.

New powerful actors, such as the American Cancer Society, took up the test and created an environment that could support the tinkering necessary to address its problems. The Pap smear faced, and faces, "chronic ambiguities" regarding the nature of cancer, the classification of normal and abnormal cells, and the reading of slides. As a result it has a false negative rate (it fails to detect cancerous and precancerous cells) of between 15 and 50 percent of cases, depending upon the circumstances. Given its initial poor reception, and the problems with even established versions of the test, what social and material adjustments allowed it to become successful, and to play a role in saving the lives of thousands, and perhaps millions, of women?

The test became less expensive by gendering the division of labor. Technicians, mainly women, could be paid less than the (predominantly male) pathologists. Technicians could even be paid on a piecework basis, and do some of their work from home. These innovations allowed the volume of testing for it to become an effective screening test. Volume, however, meant that technicians suffered from eyestrain, and from the combination of low status and high levels of responsibility on matters of life and death. Meanwhile, automation of record keeping helped to reduce the cost of Pap smears, and also to make them more useful as screening tests for large populations of women (Singleton and Michael 1993). High rates of incorrect readings of the tests have created public pressure for rating and regulating the labs performing them. Women's health activists have prompted governments to take seriously the conditions under which Pap smears are read, and the number of smears read by a technician each day, reducing the number of "Pap mills." High rates of incorrect readings have also prompted the negotiation of local orders: Rather than strictly following standard classification schemes, labs sometimes work closely with physicians and clinics, receiving information about the health of the women whose smears they read. Out of this information they develop local techniques and classifications, which, in clinical tests, appear to have better success rates.

Many material and social aspects of the test had to be constructed in order for it to be a success.

combining the construction of accounts and social reality and phenomena and the broader environment.

Many contributions to STS have converged on this point. Science and social order are "co-produced" (Jasanoff 2005). The criteria for good climate science are shaped by policy concerns, and the criteria for good policy are being shaped by climate science, in a process of "mutual construction" (Shackley and Wynne 1995). In a different way, ideas and practices concerning health and disease are bent around standardized classifications of diseases and medical interventions (Bowker and Star 2000). Assisted reproductive technologies seem miraculous, but their miracles are performed through much mundane work that integrates political, social, legal, ethical, bureaucratic, medical, technical, and quintessentially personal domains. Patients, eggs, sperm, equipment, are "choreographed" to make pregnancies, babies and parents (Thompson 2005).

6 The construction of kinds

It has been a longstanding philosophical question whether natural kinds are part of the non-human world or are only part of human classification. (The debate is usually put in terms of "universals," the abstractions that range over individual objects, like redness.) *Nominalists* believe that kinds are human impositions, even if people find it relatively easy to classify objects similarly. *Realists* believe that kinds are real features of the world, even if their edges may be fuzzy and their application somewhat conventional. For nominalists, individual objects are the only real things. Given how difficult it is to make sense of the reality of general properties – Where do they exist? How do they apply to individual objects – nominalists prefer to see them as entirely mental and linguistic phenomena. For realists, the world has to contain more than mere concrete objects. Given how difficult it is to make sense of a world without real general divisions – Do we not discover features of the world? If they were not real, why would kinds have any value? – realists see kinds as external to people.

In STS, nominalism is one way of cashing out the construction metaphor. If kinds are not features of the world, then they are constructed. To the extent that they are constructed by groups of people, they are socially constructed. Since science is the most influential institution that classifies things, science is central to the social construction of reality. We can see statements of this in works by Thomas Kuhn (e.g. 1977) and proponents of the strong programme (e.g. Barnes, Bloor, and Henry 1996): sociological finitism (Chapter 5) could be considered a version of nominalism.

7   The construction of nature

It is only a short step from nominalism to the strongest form of the social construction metaphor in STS, the claim that representations directly shape their objects. According to this form of constructivism, when scientists agree on a claim, they literally make the claim true. The world corresponds to agreement, not the other way around. Similarly, when engineers create agreement on what the most efficient solution to a problem is, they literally make that solution the most efficient one. Mind, in this case a social version of mind, is prior to nature; the way it classifies and otherwise describes the world becomes literally true. The position bears some similarity to Kant's idea that humans impose some structures on the world, so we can call it *neo-Kantian constructivism*. This constructivism has been called "the spontaneous philosophy of the critics," its popularity stemming from, among other things, the lack of institutional power of the social sciences and humanities (Guillory 2002).

Neo-Kantian constructivism gains its plausibility from two facts. First is the fact that the natures of things are not directly available without representations, that there is no independent access to the way the world is. When scientists, or other people, agree about something, they do so only in response to sense experiences, more mediated information, and arguments. Since even sense experiences are themselves responses to things, there is never direct access to the natures of things.

Second, detailed studies of science and technology suggest that there is a large amount of contingency in our knowledge before it stabilizes. Disagreement is the rule, not the exception. Yet natural and technological objects are relatively well behaved once scientists and engineers come to some agreement about them. Einstein says, "Science as something already in existence, already completed, is the most objective, impersonal thing that we humans know. Science as something coming into being, as a goal, is just as subjectively psychologically conditioned as are all other human endeavors" (quoted in Kevles 1998). We need to account for this change.

Because of the contingency of representations, we cannot say that there is a way the world is that guarantees how it will be represented. And therefore, we might question the priority of objects over their representations. Steve Woolgar probably comes closest to the neo-Kantian position in STS, when he argues that contingency undermines realist assumptions:

> The existence and character of a discovered object is a different animal according to the constituency of different social networks. . . . Crucially, this variation undermines the standard presumption about the existence of the object

prior to its discovery. The argument is not just that social networks mediate between the object and observational work done by participants. Rather, the social network constitutes the object (or lack of it). The implication for our main argument is the inversion of the presumed relationship between representation and object; the representation gives rise to the object. (Woolgar 1988: 65)

Neo-Kantian constructivism is difficult to accept. The modest version of its central claim that was put forward by Immanuel Kant was attached to an individualist epistemology: individuals impose structure on the world as they apprehend it. For the Kantian, if the individual is isolated from the material world then it makes no sense to talk of anything lying beyond phenomena, which are in part constituted by people's frameworks and preconceptions. However, STS's neo-Kantianism should not be so modest, because STS emphasizes the *social* character of scientific knowledge. If what is at issue are the representations made by groups of people, the neo-Kantian claim appears less motivated. How does consensus affect material reality? Or how do the convictions of authorities carry weight with the world that the convictions of non-authorities do not? How does it *cause* changes in the material world?

If neo-Kantian constructivism were true, for example, then representation would act at a distance without any mediators. Successful representation would change, or even create, what it represents, even though there are no causal connections from representation to represented. Neo-Kantian constructivism therefore violates some fundamental assumptions about cause and effect. For this reason authors like Latour argue that social constructivism is implausible "for more than a second" (Latour 1990).

There are also political concerns about neo-Kantian constructivism. Feminists point out (Chapter 7) that science's images of women are sometimes sexist, particularly in that they are quick to naturalize gender. If neo-Kantian constructivism were right, then, while feminist critics could attempt to change science's constructions of women, they could not reject them as false – scientific consensus is by definition true. Similarly, environmentalists have a stake in the reality of nature aside from constructions of it. While they can attempt to change dominant views on the resilience of nature, they could not reject them as false (Grundman and Stehr 2000; Takacs 1996). It may also be that constructivism places too much emphasis on contingency and the social processes of knowledge creation (Crist 2004). Scientific knowledge, the meaning of nature, environmental values, and even "natural" spaces may be shaped socially, but they are also shaped by nature.

Box 6.4   Constructivism and environmental politics

Within environmental studies, some people argue that constructivism's focus on science's social processes tends to undercut scientific knowledge (e.g. Crist 2004; Soulé and Lease 1995). STS shows that other circumstances would have produced different knowledge, yet trust in science is based on an image of science as having formulaic methods for uncovering truths of nature. Constructivism, then, appears to cripple the ability of science to serve as a solid foundation for environmental politics.

Yet that may be a misdiagnosis. Environmental politics often pits experts against each other. Experts typically try to present their own views as entirely constrained by nature and rationality, so that there is no room for disagreement. Yet those same experts find ways in which opposing arguments are open to challenge (see Chapter 11). The fact that scientific knowledge is laden with choices is not hidden, seen only by people working in STS, but is regularly rediscovered in disputes (e.g. Demeritt, 2001).

If this is right, then for science to play a larger role in politics, its knowledge should be constructed with controversy already in mind. Science's authority should not depend heavily on an incorrect formal picture of itself, at the risk of being rejected when that picture proves wrong. The constructivist view brings to the fore the complexity of real-world science, and therefore can potentially contribute to its public success. Successful science in the public sphere can be the result of the co-production of science and politics. Science can more easily solve problems in the public domain if scientific knowledge is carefully adjusted to its public contexts, and attuned to the different knowledge of others (see Chapter 16).

However, in light of the obvious truth of at least some of the other versions of constructivism described above, neo-Kantian constructivism may be a decent approximation, and may be methodologically valuable, even if it is wrong in metaphysical terms. So while claims about the "social construction of reality" can sometimes look suspect, they may amount to little more than metaphor or sweeping language. Even the political problems with neo-Kantianism may be unimportant, if the language is understood correctly (Burningham and Cooper 1999). Claims about the "social construction of reality" may draw attention to contingency in science and technology, and therefore lead researchers to ask about the causes of contingency. As a metaphor, this strong neo-Kantianism can be a valuable tool.

Richness in Diversity

At the same time that the term became common in STS, social construction talk took off in the humanities and social sciences in general, so much so that the philosopher, Ian Hacking, asks in the title of a book, *The Social Construction of What?* (Hacking 1999). Genders, emotions, identities, and political movements are only a few of the things to which social construction talk has been applied.

STS is partly responsible for this explosion of social construction talk. Because scientific knowledge is usually seen as simply reflecting the natural world, and scientists must therefore be relatively passive in the creation of that knowledge, the claim that scientific realities are socially constructed is a radical one. As a result, STS's constructivist claims have been influential. This can be seen in the explicit use of constructivist texts and ideas from STS in such fields as psychology, geography, environmental studies, education, management, cultural studies, and even accounting.

However, the diversity of claims about the social construction of reality means that constructivism in STS cannot be any neat theoretical picture. Instead, it reminds us that science and technology are social, that they are active, and that they do not take nature as it comes.

Some of the above forms of constructivism are controversial in principle, and all of them are potentially controversial in the details of their application. But given their diversity it is also clear that even the staunchest of realists cannot dismiss constructivist claims out of hand. Constructivism is the study of how scientists and technologists build socially situated knowledges and things. Such studies can even show how scientists build good representations of the material world, in a perfectly ordinary sense. As recognized by some of the different strains of constructivism, science gains power from, among other things, its ability to manipulate nature and measure nature's reactions, and its ability to translate those measurements across time and space to other laboratories and other contexts. Laboratory and other technologies thus contribute to objectivity and objective knowledge. Constructivist STS may even *support* a version of realism, then, though not the idea that there is unmediated knowledge of reality, nor the idea there is a single complete set of truths.

7

Feminist Epistemologies of Science

Can There Be a Feminist Science and Technology?

Science and engineering are not fully open to participation by women, as we saw in Chapter 4. Women face difficulties entering the education end of the pipeline, remaining there to build and maintain careers, and developing their reputation to gain status and prestige. As a result, they are under-represented, especially outside of the life sciences. In Western countries some minority groups are similarly under-represented, and outside a handful of technically highly developed countries people face even more substantial barriers to participation.

What are the effects of those barriers on scientific knowledge and technological artifacts? What would result from removing those barriers? Chapter 4 focused purely on questions of equity, rooted in issues of justice and efficiency: discrimination is clearly unjust, and is inefficient because it reduces the pool of potential contributors to science and technology. But in what ways would science and technology be qualitatively different if women were better represented? In what ways would science and technology be qualitatively different if feminist viewpoints were better represented? We might also ask in what ways science and technology would be different if Western minorities, and non-Westerners, were better represented. This chapter sketches some research relevant to these questions, and outlines of some different answers. It focuses on feminist STS, which has developed a rich and diverse body of literature.

Until recently, it would have been straightforward to argue that, were women fully represented, the content of science and engineering would not be much changed. Scientific and technical arguments would be the same, and so would be the knowledge and artifacts that they produce. If scientific knowledge simply reflects the structure of the natural world then science would not be different. If technology fills needs as efficiently as possible, given the available resources and knowledge, then it, too, would not be different. However, given the argument of this book so far, science and technology are not such purely teleological activities. What counts as knowledge and what comes to be made depend on many social and historical factors. Therefore, we should expect that feminist science and technology would be different from current science and technology. In other words, science and technology have a "political unconscious," to use Fredric Jameson's phrase, used in this context by Sandra Harding (2006).

The Technoscientific Construction of Gender

Gender is the subject of a considerable amount of scientific study, and related to a considerable amount of technological effort. For biologists and psychologists, sex-linked differences have often been important and interesting; they are, for example, central to the field of sociobiology. Whether those differences are studied in rats or in humans, differences between men and women represent at least part of the interest in the study. As a result, when it was not investigating and challenging perceptions of the place of women in science and technology, early feminist STS was almost entirely devoted to questions about the scientific construction of gender, and in particular paid attention to biologists' studies of sex and gender. Books with titles such as *Women Look at Biology Looking at Women* (Hubbard, Henifin, and Fried 1979), *Alice Through the Microscope* (Brighton Women in Science Group 1980), and *Science and Gender: A Critique of Biology and Its Theories on Women* (Bleier 1984) made important contributions to understanding science's depiction of gender, with an eye to challenging it. Many of the most prominent critics of biology were themselves biologists, challenging the research of their colleagues with the aim of improving the quality of their research.

Anne Fausto-Sterling's *Myths of Gender* (1985), for example, is a sustained attack on sex differences research. It aims to uncover poor assumptions and unwarranted conclusions, to show how researchers fail to understand the social contexts that can produce gendered behavior, and to challenge the ideological framework that supports sex difference research. Given the public interest in "biologizing" gender differences – and thus naturalizing and even legitimating them – critiques like those of Fausto-Sterling and other feminist researchers have had some urgency. The size of their impact is difficult to gauge, especially since sex differences research continues to be popular both within science and in the popular media.

Much of the study of the scientific construction of gender looks at how cultural assumptions are embedded in the language of biology. For example, since the early twentieth century, the sperm has usually been depicted as active, and the egg as passive. The egg "does not move or journey, but passively 'is transported', 'is swept', or even 'drifts' along the fallopian tube. In contrast, sperm . . . 'deliver' their genes to the egg, 'activate the developmental program of the egg', and have a 'velocity' that is often remarked on" (Martin 1991: 489; see also Biology and Gender Study Group 1989). Then, when sperm meets egg, a similar active/passive vocabulary is used, the sperm "penetrating" a "waiting" egg. This active/passive vocabulary survives despite evidence that both egg and sperm act during fertilization. Science has constructed a "romance" based on stereotypical roles, a romance that might reinforce those roles (Martin 1991).

A number of historical studies have made similar observations about depictions of males and females and their organs in biology, psychology, medicine, and anthropology. For example, mammals are called "mammals" because of the social and symbolic significance of breasts in eighteenth-century Europe – breasts feature prominently in contemporary iconography, symbols of nurturing and motherhood (Schiebinger 1993). Linnaeus introduced *Mammalia* over the previous term *Quadrupedia* in 1758, in a context that included campaigns against wet-nursing. The label served to make breast-feeding a defining natural feature of humans (and other quadrupeds), and therefore served as an argument against hiring wet-nurses.

Science can be *used*, quite consciously, to shape gender and structures that affect it. Magnus Hirschfeld, a German physician active at the turn of the twentieth century, collaborated with Austrian physiologist Eugen Steinach to argue that homosexuality was a condition rooted in biology, rather than an environmentally borne disease (Sengoopta 1998). Steinach's experiments to feminize male guinea pigs by the transplantation of ovaries provided Hirschfeld with scientific backing for his efforts to describe male homosexuality as intermediate between masculinity and femininity, and to advocate the legalization of same-sex sexual contact.

At the same time that science constructs images of gender, technology often embodies images of gender, and in so doing creates social constraints (e.g. Wajcman 1991). Built into reproductive technologies are norms of sexual behavior, desires, and families (e.g. Ginsburg and Rapp 1995). Built into domestic technologies are norms of households, standards, and divisions of labor (Cowan 1983). And built into material environments more widely are norms of divisions of labor and patterns of behavior – for example, aircraft cockpit design reflects images of users (Weber 1997).

A classic study of technology and gender is Cynthia Cockburn's *Brothers* (1983), which looks at conflicts over computerization in the London newspaper industry. In these cases, choices about technologies are the result of struggles between employers interested in reducing costs or increasing dominance, and skilled male workers interested in maintaining pay and status – particularly masculine status that comes from differentiating their work from women's work. Male industrial workers often want to work with heavier tools and machinery, making their work dependent on strength; alternatively, they may be willing to move to more managerial positions or to be put in charge of machines (Cockburn 1985). Employers, however, are often interested in the mechanization of tasks, either because it simply increases efficiency or because it allows them to hire less skilled and lower-paid workers, such as women. Mechanization and feminization, then, often go hand in hand. Industrial technologies can be deeply gendered, and have an impact on gender structures.

Interestingly, men appear more likely than women to have a particularly intimate relationship with technologies. From engineers to hobbyists, men often have love affairs with their favorite technologies (Edwards 1997; Turkle 1984). Men more than women seem to gain pleasure from the mastery and control that they have when they are skilled as makers or users of machines: "I can make this thing dance!" (Kleif and Faulkner 2003). Work and play become combined: "I like interacting with people but . . . I don't get the same kind of thrill from any other part of my job other than when I'm . . . putting my code on the machine and seeing it work" (Kleif and Faulkner 2003).

Technologies are political, because they enable and constrain action. Therefore, assumptions about gender roles that are built into technologies can, like assumptions about gender built into scientific theories, reinforce existing gender structures.

From Feminist Empiricism to Standpoint Theory

According to *feminist empiricists*, even though there are many instances of sexism in science, systematically applied scientific methods are enough to eliminate sexist science. Feminist empiricists believe that it is possible to separate a purified science from the distorting effects of society. In some sense, feminist empiricism is the position on which other feminist approaches to the sciences are grounded. It originates in the self-conception of the sciences, beginning before the articulation of other feminist epistemological positions. Yet feminist scientists and researchers in STS have been so successful at identifying sexist assumptions in scientific descriptions and theories that they have thrown doubt on the ability of the normal application of scientific methods to eliminate sexism (Harding 1986).

Helen Longino's *Science as Social Knowledge* (1990; also Longino 2002) is an attempt to explain how social ideology and the larger social context can play an important role in scientific inquiry, without abandoning objectivity and room for specific criticism in science. Longino's goal is to show the role of values in science, while leaving room to criticize the justifications of specific claims. She focuses on background assumptions: what a fact is evidence for depends on what background assumptions are held. This allows people to agree on facts and yet disagree about the conclusions to be drawn from them. At the same time, which background assumptions people choose, and which ones they choose to question, will be strongly informed by social values. Feminist scientists over the past 30 years, with experience of sexism and anti-sexist movements, approached their subject matter with novel presuppositions; the result has been a body of feminist science, especially in the biological sciences.

Although it might appear obvious, feminist empiricists have to explain why this critical work has been done primarily by women, and primarily by women sensitized to feminist issues. But to do so means rejecting purely atomistic ideas of scientists, in which they are for theoretical purposes identical, moved only by the laboratory data. No general project can isolate science from distorting social assumptions, though one can improve the assumptions that science employs. Feminist empiricism thus undermines itself, needing at least a supplement to account for its own central cases.

Feminist empiricism's trajectory naturally takes it toward *standpoint theory* (or standpoint epistemology), a theory of the privilege that particular perspectives can generate. A feminist standpoint is a *privileged* perspective, not merely *another* perspective (Hartsock 1983; Smith 1987; Rose 1986; Harding 1991; Sismondo 1995). The central argument of standpoint theory is that women's experience of sexual discrimination allows them to better understand gender relations. They are able to see aspects of discrimination that cannot be seen from the male perspective. This privileged position becomes fully available when women are active in trying to overturn male discrimination, for then they necessarily see genders as non-natural, and unjust: "A standpoint is not simply an interested position (interested as bias) but is interested in the sense of being engaged. . . . A standpoint . . . carries with it the contention that there are some perspectives on society from which, however well-intentioned one may be, the real relations of humans with each other and with the natural world are not visible" (Hartsock 1983: 285). For the task of recognizing bias and discrimination, whether in scientific theories, science, or some larger society, women are well positioned. More generally, people for whom social constraints are oppressive can more easily understand those constraints than can others.

Feminist scholars have argued that because scientific communities have lacked diversity, they have typically lacked some of the resources to better enable them to see aspects of sexism in scientific work. Therefore, standpoint theory, together with the recognition of the social character of knowledge, shows that to increase objectivity, communities of research and inquiry should be diverse, representative, and democratic.

From Difference Feminism to Anti-Essentialism

What we might call *difference feminism* claims that there are masculine and feminine perspectives and styles of knowing, which can sometimes be mapped onto those of men and women. This is different from standpoint theory's claim that social positions can offer particular insights. In STS, difference feminism is probably most associated with Evelyn Fox Keller's book *A Feeling for the Organism* (Keller 1983; also Keller 1985), a biography of Nobel Prize-winning geneticist Barbara McClintock, but can also be prominently seen in such works as Carolyn Merchant's essay on reductionism and ecology, *The Death of Nature* (1980), and Sherry Turkle's studies of gendered approaches to and uses of computers (Turkle 1984; Turkle and Papert 1990).

The central claims of difference feminism in STS revolve around the idea that there are distinct gendered styles of scientific thought: masculine knowledge is characterized by reductionism, distanced objectivity, and a goal of technical control, and feminine knowledge by attention to relationships, an intimacy between observer and observed, and a goal of holistic understanding. If this rough dichotomy is right, scientific knowledge is gendered.

The masculine and feminine are not men's knowledge and women's knowledge respectively. Instead there is a relationship between *gender* and scientific knowledge. For example, the scientist/nature relation may be coded as a male/female one, and a stereotypical key to the relation between the scientist and nature is domination; this can be seen in uses of metaphors of domination and control, rape and marriage, where (male) scientists dominate, control, rape, or marry a (female) nature (Keller 1985; Merchant 1980; Leiss 1972). The feminine approach can be represented by Keller's McClintock: "a feeling for the organism" is an appropriate slogan for the way that McClintock does genetics; rather than trying to dissect her organism she wants to understand what it is to be inside it. She says, "I found that the more I worked with them, the bigger and bigger [the chromosomes] got, and when I was really working with them I wasn't outside, I was down there. I was part of the system" (Keller 1985: 165). Her feeling for the organism is an empathetic one, and not just expertise about it.

The rough dichotomy that difference feminism posits has a particularly strong foothold in thinking about technology. Men's and women's approaches to computers can be roughly characterized in terms of a dichotomy between "hard" and "soft" mastery (Turkle 1984). Many engineers themselves employ a similar framework, understanding (masculine) technical prowess to be at the core of engineering (Hacker 1990), and thereby downplaying (feminine) social aspects of the profession. Even researchers who are cautious about the value of the dichotomy find that it has some applicability. Knut Sørensen and co-workers (Sørensen 1992; Sørensen and Berg 1987) find evidence for and against claims of gender divisions; although Sørensen finds few significant differences between male and female engineers with respect to their work, women tend to resist an eroticization of the technologies on which they are working.

Wendy Faulkner (2000) points out some interesting difficulties in the application of difference feminism to engineering. While dualisms that describe gendered approaches, styles, epistemic stances, or attitudes toward material are constant touchstones in thinking about engineering practice, they oversimplify. There are a number of related difficulties: In practice, both sides of these dualisms are necessary, and "coexist in tension"; the two sides are not valued equally, and therefore aspects of technological practice falling under the less-valued side are downplayed; and these dualisms are often gendered in complex and contradictory ways (Faulkner 2000: 760).

For example, the abstract/concrete contrast can be gendered in various ways. The ideology of engineering emphasizes concrete hands-on abilities, and those concrete abilities can be valued as masculine. Yet, engineering values the emotional detachment of mathematics, and when it does so, abstract thinking can be valued as masculine. Men can be "hands-on tinkerers," or women can be perceived as engaging in a "bricolage" style of engineering. And in practice, it may be difficult to sort out such styles from the context of the how particular engineering cultures address particular problems.

The question about gendered dichotomies in practice is quite general. Women do not *report* the same level of fascination with technology as do men, although in practice they show the same level of absorption in technicalities. Companies in which people skills are more highly valued relative to technical skills tend to see more men in roles defined around those people skills. "Men are more likely to gravitate to those roles which carry higher status (or *vice versa*)" (Faulkner 2000: 764). So, gender may be more prominent in discourse than it is in practice.

In a direct critique of Keller's work, Evelleen Richards and John Schuster (1989) make some similar points, showing that the gender structures of scientific practice are open to flexible interpretation. Looking at some variations on the story of Rosalind Franklin, whose empirical research was important to James Watson and Francis Crick's model of the structure of DNA, Richards and Schuster show that the same methods can be presented as feminine or masculine. What is gendered is talk about methods, not the methods themselves.

Such anti-essentialist challenges are tied in with movements in feminism more generally (e.g. Fraser and Nicholson 1990). One of the starting points for Donna Haraway's well-known essay, "A Manifesto for Cyborgs: Science, Technology, and Socialist Feminism in the 1980s" (1985), is the splintering of feminism in the early 1980s, when women of color made clear their frustrations with white feminism. For Haraway, the experiences and claims of women of color show that "woman" is not a completely natural political category. Identities are "fractured," structured by a multiplicity of causes. Thus the idea of discrete standpoints is open to serious question.

For Haraway, feminism should cautiously embrace science and technology, in order to play a role in shaping it around particular interests. As Haraway says, she would rather be a cyborg than a goddess. The cyborg as a unit in social theory is creatively adopted from science fiction, and emerging from a questioning of traditional political categories. This idea of the cyborg has origins in the blurring of some traditional borders or boundaries – between organism and machine, between human and animal, and between physical and non-physical – accomplished by science and technology. The cyborg might be seen as "posthuman" in its rejection of the idea that the bounded individual human is the important locus of thought and action (Hayles 1999).

Gender, Sex, and Cultures of Science and Technology

As the field has grown, feminist STS has deepened the understanding of the places of gender in science and technology, and the relations of women to science and technology. Empirical topics continually arise: science, technology, and gender are central features of the modern world, and their interactions are various, resisting uniform or synthetic treatment. Thus, issues of gender reappear in various places in this book. Gender is a category of analysis that can be and has been applied almost anywhere in STS.

Box 7.1   Masculinity in science

Gender is not just about women: questions about masculinity are interesting, too. These make, for example, science's having been an almost exclusively male province potentially interesting. One can trace male dominance in science back to the modeling of natural philosophical communities on monastic orders (Noble 1992). But there are also local explanations. Why, for instance, was the Accademia dei Lincei, an important organization of natural philosophers in Italy in the early seventeenth century, established as an exclusively male order, complete with rules against various interactions with women, and encouragements of brotherly love? The justification could have been Platonic, based on an attempt to orient the natural philosophers toward the ideal world. However, the reason appears more simply based in the misogynistic vision of Federico Cesi, who funded and controlled the Accademia, and who saw women as a distraction from the business of finding out about the natural world (Biagioli 1995).

8

Actor-Network Theory

Actor-Network Theory: Relational Materialism

Actor-network theory (ANT) is the name given to a framework originally developed by Michel Callon (e.g. 1986), Bruno Latour (e.g. 1987), and John Law (e.g. 1987). ANT has its origins in an attempt to understand science and technology, or rather *technoscience*, since on this account science and technology involve importantly similar processes (Latour 1987). ANT is, though, a general social theory centered on technoscience, rather than just a theory of technoscience.

ANT represents technoscience as the creation of larger and stronger networks. Just as a political actor assembles alliances that allow him or her to maintain power, so do scientists and engineers. However, the actors of ANT are heterogeneous in that they include both human and non-human entities, with no methodologically significant distinction between them. Both humans and non-humans form *associations*, linking with other actors to form networks. Both humans and non-humans have *interests* that cause them to act, that need to be accommodated, and that can be managed and used. Electrons, elections, and everything in between contribute to the building of networks.

Michel Callon (1987), for example, describes the effort of a group of engineers at Electricité de France (EDF) to introduce an electric car in France. EDF's engineers acted as "engineer-sociologists" in the sense that they articulated a vision simultaneously of fuel cells for these new cars, of French society into which electric cars would later fit, and of much between the two – engineering is never complete if it stops at the obvious boundaries of engineered artifacts. The EDF actors were not alone, though; their opponents at Renault, who were committed to internal combustion engines, criticized both the technical details and the social feasibility of EDF's plans, and so were also doing engineering-sociology. The engineering and the sociology are inseparable. Neither the technical vision nor the social vision will come into being without the other, though with enough concerted effort both may be brought into being together.

ANT's sociology, and the implicit sociology of the scientists and engineers being studied, deals with concrete actors rather than macro-level forces. Latour describes the efforts of the engineer Rudolf Diesel to build an earlier (than EDF's) new type of engine: "At the start, Diesel ties the fate of his engine to that of *any* fuel, thinking that they would all ignite at a very high pressure. . . . But then, nothing happened. Not every fuel ignited. This ally, which he had expected to be unproblematic and faithful, betrayed him. Only kerosene ignited, and then only erratically. . . . So what is happening? Diesel has to *shift his system of alliances*" (Latour 1987: 123). Diesel's alliances include entities as diverse as kerosene, pumps, other scientists and engineers, financiers and entre- preneurs, and consumers. The technoscientist needs to remain constantly aware of a shifting array of dramatically different actors in order to succeed. A stable network, and a successful piece of technoscience, is the result of managing all of these actors and their associations so that they contribute toward a goal.

Actors build networks. These networks might make machines function, when their components are made to act together to achieve a consistent effect. Or, they might turn beliefs into taken-for-granted facts, when their com- ponents are made to act as if they are in agreement. So working machines and accepted facts are the products of networks. The activity of technoscience, then, is the work of understanding the interests of a variety of actors, and *translating* those interests so that the actors work in agreement (Callon 1986; Callon and Law 1989). That is, in order to form part of a network, an actor must be brought to bear on other actors, so they must be brought together. Moreover, they must be brought together so as to work together, which may mean changing the ways in which they act. By being moved and changed, interests are translated in both place and form. In this way, actors are made to act; as originally defined, the actors of ANT are *actants*, things *made to act*.

ANT is a materialist theory. It reduces even the "social" to the material, both inside and outside of science (Latour 2005). Science and technology work by translating material actions and forces from one form into another. Scientific representations are the result of material manipulations, and are solid precisely to the extent that they are mechanized. The rigidity of trans- lations is key here. Data, for example, is valued as a form of representation because it is supposed to be the direct result of interactions with the natural world. Visiting an ecological field site in Brazil, Latour (1999) observes researchers creating data on the colors of soil samples. So that the color of the sample can be translated into a uniform code, Munsell color charts are held against the samples (just as a painter will hold a color chart against a paint sample). As Latour jokes, the gap between representation and the world, a standard philosophical problem that gives rise to questions about realism (Chapter 6), is reduced by scientists to a few millimeters. Data-level representations are themselves juxtaposed to form new relationships that are summarized and otherwise manipulated to form higher-level representations, representations that are more general and further from their objects. Again, the translation metaphor is apt, because these operations can be seen as translations of representations into new forms, in which they will be more generally applicable. Ideally, there should be no leaps between data and theory – and between theory and application – but only a series of minute steps. There is no action at a distance, though through the many translations or linkages there may be long-distance control (see Star 1989).

Again, science and technology must work by translating material actions and forces from one form into another. The working of abstract theories and other general knowledge appears a miracle unless it can systematically be derived from or traced to local interactions, via hands-on manipulation and working machines, via extractions from original settings, via data, and via techniques for summarizing, grouping, and otherwise exploiting information. This is the methodological value of materialism. Universal scientific knowledge is the product of the manipulation of local accounts, a product that can supposedly be transported through time and space to a wide variety of new local circumstances. But such universal knowledge is only applicable through a new set of manipulations that adapt it once again to those local circumstances (or adapt those local circumstances to it). Sciences have to solve the problem of action at a distance, but in so doing they work toward a kind of universality of knowledge.

Seen in these terms, laboratories give scientists and engineers power that other people do not have, for "it is in the laboratories that most new sources of power are generated" (Latour 1983: 160). The laboratory contains tools, like microscopes and telescopes, that change the effective sizes of things. Such tools make objects human in scale, and hence easier to observe and manipulate. The laboratory also contains a seemingly endless variety of tools for separating parts of objects, for controlling them, and for subjecting them to tests: objects are tested to find out what they can and cannot do. This process can also be thought of as a series of tests of actors, to find out which alliances can and cannot be built. Simple tools like centrifuges, vacuum pumps, furnaces, and scales have populated laboratories for hundreds of years; these and their modern descendents tease apart, stabilize, and then quantify objects, enabling a kind of *engine science* (Carroll 2006). *Inscription devices*, or machines that "transform pieces of matter into written documents" (Latour and Woolgar 1986: 51), allow researchers to deal with nature on pieces of paper. Like the representations produced by telescopes and

Box 8.1   The Pasteurization of France

Louis Pasteur's anthrax vaccine is the subject of an early statement of actor-network theory (Latour 1983), and Pasteur's broader campaign on the microbial theory of disease is the subject of a short book (Latour 1988).

How could Pasteur be seen as the central cause of a revolution in medicine and public health, even though he, as a single actor, could do almost nothing by himself? The laboratory was probably the most important starting point. Here is Pasteur, describing the power of the laboratory:

> As soon as the physicist and chemist leave their laboratories, . . . they become incapable of the slightest discovery. The boldest conceptions, the most legitimate speculations, take on body and soul only when they are consecrated by observation and experience. Laboratory and discovery are correlative terms. Eliminate the laboratories and the physical sciences will become the image of sterility and death . . . Outside the laboratories, the physicists and chemists are unarmed soldiers in the battlefield. (in Latour 1988: 73)

Pasteur used the strengths of the laboratory to get microbes to do what he wanted. Whereas in nature microbes hide, being invisible components of messy constellations, in the laboratory they could be isolated and nurtured, allowing Pasteur and his assistants to deal with visible colonies. These could be tested, or subjected to *trials of strength*, to find their properties. In the case of microbes, Pasteur was particularly interested in finding weak versions that could serve as vaccines.

Out of the complex set of symptoms and circumstances that make a disease, Pasteur *defined* a microbe in the laboratory; his manipulations and records specified its boundaries and properties. He then was able to argue, to the wider scientific and medical community, that his microbe was responsible for the disease. This was in part done via public demonstrations that repeat laboratory experiments. Breakthroughs like the successful vaccination of sheep against anthrax were performed in carefully staged demonstrations, in which the field was turned into a laboratory, and the public was invited to witness the outcomes of already-performed experiments. Public demonstrations helped convince people of two important things: that microbes are key to their goals, whether those goals are health, the strength of armies, or public order; and that Pasteur had control over those microbes.

Microbes were not merely entities that Pasteur studied, but agents with whom Pasteur built an alliance. The alliance was ultimately very successful.

It created enormous interest in Pasteur's methods of inquiry, reshaped public health measures, and brought prestige and power to Pasteur. We might see Pasteur's work as having introduced a new element into society, an element of which other people have to take account if they are to achieve their goals.

When doctors, hygienists, regulators, and others put in place measures oriented around Pasteur's purified microbe, it became a taken-for-granted truth that the microbe was the real cause of the disease, and that Pasteur was the cause of a revolution in medicine and public health.

---

microscopes these are also medium-sized, but perhaps more importantly they are durable, transportable, and relatively easy to compare to each other. Such *immutable mobiles* can be circulated and manipulated independently of the contexts from which they derive. Nature brought to a human scale, teased into components, made stable in the laboratory, and turned into marks on paper or in a computer, is manipulable, and manipulable at leisure in *centers of calculation* (Latour 1987) where inscriptions can be combined and analyzed to produce abstract and general representations. When they become accepted, those representations are often taken to be Nature, rather than products of or interpretations of nature.

We can see that, while ANT is a general theory, it is one that explains the centrality of science and technology to the idea of modernity (Latour 1993). Technologies reshape the field of agency, because people delegate agency to them. Science and technology explicitly engage in crossing back and forth between objects and representations, creating more situations in which humans and non-humans affect each other. Science and technology are responsible for the contemporary world, because more than any other activities they have mixed humans and non-humans together, allowing a dramatic expansion of the social world. Science and technology have brought non-humans into the human world, to shape, replace, and enlarge social organizations, and have brought human meanings and organizations to the non-human world, to create new alignments of forces (Latour 1994).

But although the material processes by which facts and machines are produced may be very complex, science and engineering's networks often stabilize and become part of the background or invisible. Configurations become *black boxes*, objects that are taken for granted as completed projects, not as messy constellations. The accumulation of black boxes is crucial for what is considered progress in science and engineering. As philosopher Alfred

North Whitehead (1992 [1911]) wrote, "Civilization advances by extending the number of important operations which we can perform without thinking about them."

While actor-network theory is thoroughly materialist, it is also built on a relational ontology; it is based on a *relational materiality* (Law 1999). Objects are defined by their places in networks, and their properties appear in the context of tests, not in isolation. Perhaps most prominently, not only technoscientific objects but also social groups are products of network-building. Social interests are not fixed and internal to actors, but are changeable external objects. The French military of 1880 was interested in recruiting better soldiers, but Louis Pasteur translated that interest, via rhetorical

---

Box 8.2   Ecological thinking

Science and technology are done in rich contexts that include material circumstances, social ties, established practices, and bodies of knowledge. Scientific and technological work is performed in complex ecological circumstances; to be successful, that work must fit into or reshape its environment.

An ecological approach to the study of science and technology emphasizes that multiple and varying elements contribute to the success of an idea or artifact – and any element in an idea or artifact's environment may be responsible for failure. An idea does not by itself solve a problem, but needs to be combined with time to develop it, skilled work to provide evidence for it, rhetorical work to make it plausible to others, and the support to put all of those in place. If some of the evidential work is empirical, then it will also demand materials, and the tinkering to make the materials behave properly. Solutions to problems, therefore, need nurturing to succeed.

There is no *a priori* ordering of such elements. That is, no one of them is crucial in advance. With enough effort, and with enough willingness to make changes elsewhere in the environment, anything can be changed, moved, or made irrelevant. (This is a generalized version of the Duhem–Quine thesis, Box 1.2.) As a result, there is no *a priori* definition of good and bad ideas or good and bad technologies. Success stories are built out of many distinct elements. They are typically the result of many different innovations, some of which might normally be considered technical, some economic, some social, and some political. The "niche" of a technological artifact or a scientific fact is a multi-dimensional development.

work, into support for his program of research. After Pasteur's work, the military had a new interest in basic research on microbes. Translation in ANT's sense is not neutral.

Whereas the strong programme was "symmetric" in its analysis of truth and falsity and in its application of the same social explanation for, say, both true and false beliefs, ANT is "supersymmetric," treating both the social and material worlds as the products of networks (Callon and Latour 1992; Callon and Law 1995). Representing both human and non-human actors, and treating them in the same relational terms, is one way of prompting full analyses, analyses that do not discriminate against any part of the ecologies of scientific facts and technological objects. It does not privilege any particular set of variables, because every variable depends upon others. Networks confront each other as wholes, and to understand their successes and failures STS has to study the wholes (and the parts) of those networks.

Some Objections to Actor-Network Theory

Actor-network theory, especially in the form articulated by Bruno Latour in his widely read book *Science in Action* (1987), has become a constant touchstone in STS, and is increasingly being exported into other domains. The theory is easy to apply to, and can offer insights on, an apparently limitless number of cases. Its focus on the materiality of relations creates research problems that can be solved, through analyses of the components and linkages of any given network. Yet its broad application of materialism, and the fact that its materialism is relational, means that its applications are often counter-intuitive. This success does not, though, mean that STS has uncritically accepted ANT. The remainder of this chapter is devoted to criticisms of the theory. This discussion of problems that ANT faces is not supposed to indicate the theory's failure, but instead should contribute to further explaining the theory and demonstrating its scope.

1   Practices and cultures

Actor-network theory, and for that matter almost every other approach in STS, portrays science and engineering as rational in a means–end sense: technoscientists use the resources that are available – rhetorical resources, established power, facts, and machines – to achieve their goals. Of course, rational choices are not made in a vacuum, or even only in a field of simple material and conceptual resources. They are made in the context of

Box 8.3   The Mathematical Tripos and the Cambridge culture of mathematical physics

For much of the nineteenth century, placing highly on Cambridge University's Mathematical Tripos exam was a mark of the successful Cambridge undergraduate, even among those not intending to go on in mathematized fields. The pressure to perform well spawned systems of private tutoring, intense study, and athletic activity, in which students mastered problems, techniques, heuristics, and bodily discipline – university athletics as a whole arose in part because of the exam (Warwick 2003). Students hoping to score among the top group had little choice but to hire the top tutors and submit to their highly regimented plans of study; those tutors, some of them brilliant at both mathematical physics and pedagogy, taught many of the most important physicists of the century. The increasing ability of the undergraduates, and competition in the relatively closed world of those setting the exam, led to the Tripos's increasing difficulty, and also fame: lists of those in the order of merit, with special attention on those placed as "Wranglers" at the top, were printed in the *Times* of London from 1825 to 1909.

All of this activity created a distinctive culture of mathematical physics at Cambridge, one centered on a particular array of skills and examination-sized problems. Even James Clerk Maxwell's 1873 *Treatise on Electricity and Magnetism*, one of the key works of physics of the nineteenth century, was partly written as a textbook for Cambridge undergraduates, and featured case-by-case solutions to problems. That difficult work, in turn, became important to pedagogy, through its consistent and careful interpretation by other physicists, at Cambridge and elsewhere in Britain. The result was a distinctive style of classical physics in Britain, that was, for example, practically incommensurable with the new styles of physics that arose in Germany in the twentieth century, because its core mathematical practices were different (Warwick 2003).

existing technoscientific cultures and practices. Practices can be thought of as the accepted patterns of action and styles of work; cultures define the scope of available resources (Pickering 1992). Opportunistic work, even work that transforms cultures and practices, is an attempt to appropriately combine and recombine cultural resources to achieve particular goals. Practices and cultures provide the context and structure for technoscientific opportunism. But because ANT treats humans and non-humans on the same footing, and because it adopts an externalized view of actors, it does not pay attention to such distinctively human and apparently subjective factors as cultures and practices.

Cultures and cultural networks do not fit neatly into the network framework offered by ANT. For example, mathematical physicists at Cambridge University developed a particular style of work and theorizing (Warwick 2003; see Box 8.3). The result was a generalized culture of physics that shaped and was shaped by pedagogy, skills, and networks. To take another example, trust is an essential feature of scientific and technological work, in that researchers rely upon findings and arguments made by people they have never met, and about whom they may know almost nothing. But trust is often established through faith in a common culture. The structure of trust in science was laid down by being transferred from the structure of gentlemanly trust in the seventeenth century; gentlemen could trust each other, and could not easily challenge each other's truthfulness (Shapin 1994). Similarly, trust in technical judgment often resides in cultural affiliations. Engineers educated in the École Polytechnique in nineteenth-century France trusted each other's judgments (Porter 1995), just as did engineers educated at the Massachusetts Institute of Technology in the twentieth century (e.g. MacKenzie 1990).

To account for even rational choices we need to invoke practices and cultures. Yet the world of ANT is culturally flat. Within the terms of the theory practices and cultures need be understood in terms of arrangements of actors that produce them. Macro-level features of the social world have to be reducible to micro-level ones, without action at a distance. While that is possibly very attractive, the reduction represents a large promissory note.

2 Problems of agency

Actor-network theory has been criticized for its distribution of agency. On the one hand, it may encourage analyses centered on key figures, and perhaps as a result many of the most prominent examples are of heroic scientists and engineers, or of failed heroes. The resulting stories miss work done by other actors, miss structures that prevent others from participating, and miss non-central perspectives. Marginal, and particularly marginalized, perspectives may provide dramatically different insights; for example, women who are sidelined from scientific or technical work may see the activities of science and technology quite differently (e.g. Star 1991). With ANT's focus on agency, positions from which it is difficult to act make for less interesting positions from which to tell stories. So ANT may encourage the following of heroes and would-be heroes.

On the other hand, actor-network analyses can be centered on any perspective, or on multiple perspectives. Michel Callon even famously uses the perspective of the scallops of St. Brieuc Bay for a portion of one important statement of ANT (Callon 1986). This positing of non-human agents is one of the more controversial features of the theory, attracting a great deal of criticism (see, e.g., Collins and Yearley 1992).

In principle, ANT is entirely symmetrical around the human/non-human divide. Non-humans can appear to act in exactly the same way as do humans – they can have interests, they can enroll others. (Because ANT's actors are *actants*, things made to act, agency is an effect of networks, not prior to them. This is a difficult distinction to sustain, and the ends of ANT's analyses seem to rest on the agency of non-humans.) Critics, though, argue that humans and non-humans are crucially different. Humans have, and most non-humans do not have, intentionality, which is necessary for action on traditional accounts of agency. To treat humans and non-humans symmetrically, ANT has to deny that intentionality is necessary for action, and thus deny that the differences between humans and non-humans are important for the theory overall.

In practice, though, actor-network analyses tend to downplay any agency that non-humans might have (e.g. Miettinen 1998). Humans appear to have richer repertoires of strategies and interests than do non-humans, and so tend to make more fruitful subjects of study. The subtitle of Latour's popular *Science in Action* is *How to Follow Scientists and Engineers through Society*, suggesting that however symmetric ANT is, of particular interest are the actions of scientists and engineers.

3   Problems of realism

Running parallel to problems of agency are problems of realism. On the one hand, ANT's relationalism would seem to turn everything into an outcome of network-building. Before their definition and public circulation through laboratory and rhetorical work, natural objects cannot be said to have any real scientific properties. Before their public circulation and use, artifacts cannot be said to have any real technical properties, to do anything. For this reason, ANT is often seen, despite protests by actor-network theorists, as a blunt version of constructivism: what is, is constructed by networks of actors. This constructivism flies in the face of strong intuitions that scientists discover, rather than help create, the properties of natural things. It flies in the face of strong intuitions that technological ideas have or lack force of their own accord, whether or not they turn out to be successful. And this constructivism runs against the arguments of realists that (at least some) things have real and intrinsic properties, no matter where in any network they sit.

On the other hand, positing non-human agents appears to commit ANT to realism. Even if ANT assumes that scientists in some sense define or construct the properties of the so-called natural world, it takes their interests seriously. That is, even if an object's interests can be manipulated, they resist that manipulation, and hence push back against the network. This type of picture assumes a reality that is prior to the work of scientists, engineers, and any other actors. Latour says, "A little bit of constructivism takes you far away from realism; a complete constructivism brings you back to it" (Latour 1990: 71).

Theorists working outside the ANT tradition are faced with similar problems. For example, Karen Barad (2007) articulates a position she calls *agential realism*: human encounters with the world take the form of *phenomena*, which are ontologically basic. Material-discursive practices create intra-actions within these phenomena. These parcel out features of the world and define them as natural or human. Similarly, Andrew Pickering's pragmatic realism (1995) describes a *mangle of practice* in which humans encounter *resistances* to which they respond. Technologies and facts about nature result from a dialectic of resistance and accommodation. Barad's and Pickering's frameworks, which share features with ANT and with each other, are designed to bridge constructivist and realist views.

The implicit realism of ANT has been both criticized, as a step backwards from the successes of methodological relativism (e.g. Collins and Yearley 1992), and praised as a way of integrating the social and natural world into STS (Sismondo 1996). For the purposes of this book, whether ANT makes realist assumptions, and whether they might move the field forwards or backwards are left as open questions, much as they have been in STS itself.

4 Problems of the stability of objects and actions

A further problem facing ANT will be made more salient in later chapters. According to the theory, the power of science and technology rests in the arrangement of actors so that they form literal and metaphorical machines, combining and multiplying their powers. That machining is made possible by the power of laboratories and laboratory-like settings (such as field sites) that are made to mimic labs (Latour 1999). As noted above, the power of laboratories depends upon material observations and manipulations that we presume to be repeatable and stable. Once an object has been defined and characterized, it can be trusted to behave similarly in all similar situations, and actions can be delegated to that object.

Science and technology gain power from the translation of forces from context to context, translations that can only be consistently achieved by formal rules. However, rules have to be interpreted, and Wittgenstein's problem of rule following shows that no statement of a rule can determine its interpretations (Box 3.2). As we will see, STS has shown how tinkering is crucial to science and technology, how the work of making observations and manipulations is difficult, how much routine science and engineering involves expert judgment, and how that judgment is not reducible to formulas (Chapter 10). ANT, while it recognizes the provisional and challengeable nature of laboratory work, glides over these issues. It presents science and technology as powerful because of the rigidity of their translations, or the objectivity – in the sense that they capture objects – of their procedures. Yet rigidity of translation may be a fiction, hiding many layers of expert judgment.

Conclusions

Especially since the publication of Latour's *Science in Action* (1987), ANT has dominated theoretical discussions in STS, and has served as a framework for an enormous number of studies. Its successes, as a theory of science, technology, and everything else, have been mostly bound up in its relational materialism. As a materialist theory it explains intuitively the successes and failures of facts and artifacts: they are the effects of the successful translation of actions, forces, and interests. As a relationalist theory it suggests novel results and promotes ecological analyses: humans and non-humans are bound up with each other, and features on neither side of that apparent divide can be understood without reference to features on the other. Whether actor-network theorists can answer all the questions people have of it remains to be seen, but it stands as the best known of STS's theoretical achievements so far.

9

Two Questions Concerning Technology

Is Technology Applied Science?

The idea that technology is applied science is now centuries old. In the early seventeenth century, Francis Bacon and René Descartes both promoted scientific research by claiming that it would produce useful technology. In the twentieth century this view was importantly championed by Vannevar Bush, one of the architects of the science policy pursued by the United States after World War II: "Basic research . . . creates the fund from which the practical applications of knowledge must be drawn. New products and new processes do not appear full-grown. They are founded on new principles and new conceptions, which in turn are painstakingly developed by research in the purest realms of science. . . . Today, it is truer than ever that basic research is the pacemaker of technological progress." The basic-applied research connection is part of a "linear model" that traces innovation from basic research to applied research to development and finally to production. That linear model developed over the first two-thirds of the twentieth century, as a rhetorical tool used by scientists, management experts, and economists (Godin 2006).

However, accounts of artifacts and technologies show that scientific knowledge plays little direct role in the development of even many state of the art technologies. Historians and other theorists of technology have argued that there are technological knowledge traditions that are independent of science, and that to understand the artifacts one needs to understand them.

Because of its large investment in basic research, in the mid-1960s the US Department of Defense conducted audits to discover how valuable that research was. Project Hindsight was a study of key events leading to the development of 20 weapons systems. It classified 91% of the key events as technological, 8.7% as applied science, and 0.3% as basic science.

Project Hindsight thus suggested that the direct influence of science on technology was very small, even within an institution that invested heavily in science and was at key forefronts of technological development. A subsequent study, TRACES, challenged that picture by looking at prominent civilian technologies and following their origins further back in the historical record.

Among historians of technology it is widely accepted that "science owes more to the steam engine than the steam engine owes to science." Science is applied technology more than technology is applied science. As we saw in the last chapter, scientific work depends crucially on tools for purifying, controlling, and manipulating objects. Meanwhile, technology may be relatively divorced from science. Work on the history of aircraft suggests that aeronautical engineers consult scientific results when they see a need to, but their work is not driven by science or the mere application of science (Vincenti 1990). Similarly, the innovative electrical engineer Charles Steinmetz did not either apply physical theory or derive his own theoretical claims from it (Kline 1992), but instead developed theoretical knowledge in purely engineering contexts. Engineers, then, develop their own mathematics, their own experimental results, and their own techniques.

Technology is so often seen as applied science because technological knowledge is downplayed (Layton 1971, 1974). In the nineteenth century, for example, American engineers developed their own theoretical works on the strength of materials, drawing on but modifying earlier scientific research. When engineers needed results that bore on their practical problems, they looked to engineering research, not pure science. Engineers and inventors participate in knowledge traditions, which shape the work that they do, especially work that fits into technological paradigms (Constant 1984). Science, then, does not have a monopoly on technical knowledge. The development of technologies is a research process, driven by interesting problems: actual and potential functional failure of current technologies, extrapolation from past technological successes, imbalances between related technologies, and, more rarely, external needs demanding a technical solution (Laudan 1984). All but the last of these problem sources stem from within technological knowledge traditions.

For a group of people to have its own tradition of knowledge means that that knowledge is tied to the group's social networks and material circumstances. As we have seen, there is some practical incommensurability between knowledge traditions, seen in the difficulties of translating between traditions. In addition, some knowledge within a tradition is tacit, not fully formalizable, and requires socialization to be passed from person to person (Chapter 10).

So far, we have seen arguments that technological practice is autonomous from science. A separate set of arguments challenge the idea that technology is applied science from almost the opposite direction. Some people working in STS have argued that science and technology are not sufficiently well defined and distinct for there to be any determinate relationship between them. In the context of large technological systems, "persons committed emotionally and intellectually to problem solving associated with system creation and development rarely take note of disciplinary boundaries, unless bureaucracy has taken command" (Hughes 1987). "Scientists" invent, and "inventors" do scientific research – whatever is necessary to move their program forward.

The indistinctness of science and technology can fall out of accounts of science, as well. First, "basic research" turns out to be a flexible and ambiguous concept, having a history and being used in different ways (Calvert 2006): Scientists use the term in order to do boundary work, drawing on the prestige of ideals of purity to gain funding and independence. Second, for the pragmatist, scientific knowledge is about what natural objects can be made to do. Thus laboratory science may be seen to be about what can be constructed, not about what exists independently (Knorr Cetina 1981). For the purposes of this chapter, the pragmatic orientation is relevant in that it draws attention to the ways in which science depends upon and involves technology, both materially and conceptually.

Actor-network theory's term *technoscience* eschews a principled conceptual distinction between science and technology. It also draws attention to the increasing causal interdependence of what is labeled science and technology. We might think it odd that historians are insisting on the autonomy of technological traditions and cultures precisely when there is a new spate of science-based technologies and technologically oriented science – biotechnologies, new materials science, and nanotechnology all cross obvious lines.

Latour's networks and Thomas Hughes's technological systems bundle many different resources together. Thomas Edison freely mixed economic calculations, the properties of materials, and sociological concerns in his designs (Hughes 1985). Technologists need scientific and technical knowledge, but they also need material, financial, social, and rhetorical resources. Even ideology can be an input, in the sense that it might shape decisions and the conditions of success and failure (e.g. Kaiserfeld 1996). For network builders nothing can be reduced to only one dimension. Technology requires heterogeneous engineering of a dramatic diversity of elements (Law 1987; Bucciarelli 1994). A better picture of technology, then, is one that incorporates many different inputs, rather than being particularly dependent upon a single stream. It is possible that no one input is even essential: any input could be worked around, given enough hard work, ingenuity, and other resources.

To sum up, scientific knowledge is a resource on which engineers and inventors can draw, and perhaps on which they are drawing increasingly, but on the whole it is not a driver of technology. Rather, technological development is a complex process that integrates different kinds of knowledge – including its own knowledge traditions – and different kinds of material resources. At the same time, science draws on technology for its instruments, and perhaps also for some of its models of knowledge, just as some engineers may draw on science for their models of engineering knowledge. There are multiple relations of science and technology, rather than a single monolithic relation. "The linear model . . . is dead" (Rosenberg 1994).

Does Technology Drive History?

Technological determinism is the view that material forces, and especially the properties of available technologies, determine social events. The reasoning behind it is usually economistic: available material resources form the environment in which rational economic choices are made. In addition, technological determinism emphasizes "real-world constraints" and "technical logics" that shape technological trajectories (Vincenti 1995). The apparent autonomy of technologies and technological systems provides some evidence of these technical logics: technologies behave differently and enter different social contexts than their inventors predict and desire. If this is right, then social variables ultimately depend upon material ones.

A few of Karl Marx and Friedrich Engels's memorable comments on the influence of technology on economics and society can stand in for the position of the technological determinist, though they are certainly not everything that Marx and Engels had to say about the determinants of social structures. Looking at large-scale structures, Marx famously said "The hand-mill gives you society with the feudal lord, the steam-mill, society with the industrial capitalist." Engels, talking about smaller-scale structures, claimed that "The automatic machinery of a big factory is much more despotic than the small capitalists who employ workers ever have been."

There are a number of different technological determinisms (see Bimber 1994; Wyatt 2007), but the central idea is that technological changes force social adaptations, and consequently constrain the trajectories of history. Robert Heilbroner, supporting Marx, says that the hand-mill (if we may take this as referring to late medieval technology in general) required a work force composed of skilled or semiskilled craftsmen, who were free to practice their occupations at home or in a small atelier, at times and seasons that varied considerably. By way of contrast, the steam-mill – that is, the technology of the nineteenth century – required a work force composed of semiskilled or unskilled operatives who could work only at the factory site and only at [a] strict time schedule. (Heilbroner 1994 [1967])

Because economic actors make rational choices, class structure is determined by the dominant technologies. This reasoning applies to both the largest scales and much more local decisions. Technology, then, shapes economic choices, and through them shapes history.

Some technologies appear compatible with particular political and social arrangements. In a well-known essay, Langdon Winner (1986a) asks "do artifacts have politics?" Following Engels, he argues that some complex technological decisions lend themselves to more hierarchical organization than others, in the name of efficiency – the complexity of modern industrial production does not sit well with consensus decision-making. In addition, Winner argues, some technologies, such as nuclear power, are dangerous enough that they may bring their own demands for policing, and other forms of state power. And finally, individual artifacts may be constructed to achieve political goals. For example, the history of industrial automation reveals many choices made to empower and disempower different key groups (Noble 1984): Numerical control automation, the dominant form, was developed to eliminate machinist skill altogether from the factory floor, and therefore to eliminate the power of key unions. While also intended to reduce factories' dependence on skilled labor, record-playback automation, a technology not developed nearly as much, would have required the maintenance of machinist skill to reproduce it in machine form (for some related issues see Wood 1982). In general, technologies are deskilling. In replacing labor they also replace the skills that are part of that labor. Even a technology like a seed can have that effect. Before hybrid corn was introduced in the 1930s, American farmers were skilled at breeding their own corn, for high yield and disease resistance (Fitzgerald 1993). Lines of hybrid seed, though, could not be continued on the farm, since they were first-generation crosses of inbred lines; this ensured that seed producers could sell seed every year, which was an explicit goal. When American farmers bought high-yield hybrid seed, which they did willingly, they were delegating their breeding work to the seed companies, and setting their breeding skills aside.

Even for non-determinists, the effects of technologies are important. As we saw in Chapter 1, a key part of the pre-STS constellation of ideas on science and technology was the study of the positive and negative effects of technologies, and the attempt to think systematically about these effects. That type of work continues. At the same time, as we see below, some STS researchers have challenged a seemingly unchallengeable assumption, the assumption that technologies have any systematic effects at all! In fact, they challenge something slightly deeper, the idea that technologies have essential features (Pinch and Bijker 1987; Grint and Woolgar 1997). If technologies have no essential features, then they should not have systematic effects, and if they do not have any systematic effects then they cannot determine structures of the social world.

No technology – and in fact no object – has only one potential use. Even something as apparently purposeful as a watch can be simultaneously constructed to tell time, to be attractive, to make profits, to refer to a well-known style of clock, to make a statement about its wearer, etc. Even the apparently simple goal of telling time might be seen a multitude of different goals: within a day one might use a watch to keep on schedule, to find out how long a bicycle ride took, to regulate the cooking of a pastry, to notice when the sun set, and so on. Given this diversity, there is no essence to a watch. And if the watch has no essence, then we can say that it has systematic effects only within a specific human environment.

In their work on "Social Construction of Technology" (SCOT), Trevor Pinch and Wiebe Bijker (1987) develop this point into a framework for thinking about the development of technologies. In their central example, the development of the safety bicycle, the basic design of most twentieth-century bicycles, there is an appearance of inevitability about the outcome. The standard modern bicycle is stable, safe, efficient, and fast, and therefore we might see its predecessors as important, but ultimately doomed, steps toward the safety bicycle. On Pinch and Bijker's analysis, though, the safety bicycle did not triumph because of an intrinsically superior design. Some users felt that other early bicycle variants represented superior designs, at least superior to the early versions of the safety bicycle with which they competed. For many young male riders, the safety bicycle sacrificed style for a claim to stability, even though new riders did not find it very stable. Young male riders formed one *relevant social group* that was not appeased by the new design. Their goals were not met by the safety bicycle, as its meaning (to them) did not correspond well to their understanding of a quality bicycle. There is, then, *interpretive flexibility* in the understanding of technologies and in their designs. We should see trajectories of technologies as the result of rhetorical operations, defining the users of artifacts, their uses, and the problems that particular designs solve. The Luddites of early-nineteenth-century Britain adopted a variety of interpretations of the factory machines that they did and did not smash (Grint and Woolgar 1997). Although some saw the factory machines as upsetting their preferred modes of work, others saw the problem in the masters of the factories. Resistance to the new technologies diminished when new left-wing political theories articulated the machines as saviors of the working classes. The machines, then, did not have a single consistent set of effects.

Interpretive flexibility can create quite unexpected results. For example, the use of "safer sex" technologies by prostitutes shows how users can give technologies novel meanings, even in the service of the ends for which it is made (Moore 1997). In order to fit the contexts and culture in which prostitutes use it, an apparently clinical latex rubber glove can become sexy by being snapped onto hands in the right way, or can be put to new uses by being cut and reconfigured. Very high numbers of users modify products for their own use, in some cases giving very new meanings to those products (von Hippel 2005).

On a SCOT analysis, the success of an artifact depends upon the strength and size of the group that takes it up and promotes it. Its definition depends upon the associations that different actors make. Interpretive flexibility is a necessary feature of artifacts, because what an artifact does and how well it performs are the results of a competition of different groups' claims. Thus the good design of an artifact cannot be an independent cause of its success; what is considered good design is instead the result of its success.

Attention to users reveals the variety of relations that users and technologies have (Oudshoorn and Pinch 2003). Among other things, users contribute to technological change, not just by adapting objects to their local needs, but also by feeding back into the design and production processes. In its early days the automobile was put to novel uses by farmers, for example, who turned it into a mobile source of power for running various pieces of farm machinery; their innovations contributed to the features that became common on tractors (Kline and Pinch 1996). In general, *lead users*, those who are early adopters of a new technology that goes on to have wide use, and who gain considerable benefits from innovating, tend to make substantial innovations that can be picked up by producers of that technology (von Hippel 2005). This poses another problem for the linear model of innovation.

A different kind of novel use of technology is the use of different *ideologies* of technology. The "electromechanical vibrator" was widely sold, and advertised in catalogues, between 1900 and 1930, despite the fact that masturbation was socially prohibited in this period, and was even thought to be a cause of hysteria (Maines 2001). The vibrator could be acceptable for sale because of its association with professional instruments, because of the high value attached to electric appliances in general, and because electricity

Box 9.1   Digital rights management

With the advent of digital culture, debates about copyright became louder and acquired new dimensions. Digital cultural objects such as CDs, DVDs, and MP3s can in principle be copied indefinitely many times without any degradation. That worries many people, most importantly people in the powerful US music and film industries. One result has been Digital Rights Management (DRM), explored by Tarleton Gillespie in his book *Wired Shut* (2007). DRM is ostensibly an attempt by copyright holders to directly prevent illegal, and some legal, copying, by encrypting files so that they may only be copied in authorized ways.

By itself DRM cannot work. The history of encryption is a history of struggles between coders and decoders. Moreover, for DRM to work for the cultural industries, the decryption codes must be very widely shared, or else sales will be very limited. "The real story, the real power behind DRM, is . . . the institutional negotiations to get technology manufacturers to build their devices to chaperone in the same way, the legitimation to get consumers to agree to this arrangement . . . and the legal wrangling to make it criminal to circumvent" (Gillespie 2007: 254). And the cultural industries in the United States have successfully convinced the US government to try to spread protection of DRM to the rest of the world through intellectual property negotiations.

But in this social context, DRM has real effects. It prevents some copying, both legal and illegal. It gives content providers more control over the uses of their products, which may allow them to find new business models on which they can charge per use, for different types of uses, and can charge differently in different markets.

was seen as a healing agent. The modernity of technology, then, could be used to revalue objects and practices.

To give technologies reasonably determinate meanings requires work. In one study, at a point well into the development process, a computer firm needed to test prototypes of their package, to see how easy it was for unskilled users to figure out how to perform some standard tasks (Grint and Woolgar 1997). On the one hand these tests could be seen as revealing what needed to be done to the computers in order for them to be more user-friendly. On the other hand they could be seen as revealing what needed to be done to the users – how they needed to be defined, educated, controlled – to make them more computer-friendly: successful technologies require what Steve Woolgar calls *configuring the user*. The computer, then, does what it does only in the context of an appropriate set of users.

Surely, though, some features of technologies are autonomous and defy interpretation. We might, for example, ask with Rob Kling (1992) "What's so social about being shot?" Everything, say Grint and Woolgar. In a tour-de-force of anti-essentialist argumentation, Grint and Woolgar argue that a gun being shot is not nearly as simple a thing as it might seem. It is clear that the act of shooting a gun is intensely meaningful – some guns are, for example, more manly than others. But more than that, even injuries by gun-shot can take on different meanings. When female Israeli soldiers were shot in 1948, "men who might have found the wounding of a male colleague comparatively tolerable were shocked by the injury of a woman, and the mission tended to get forgotten in a general scramble to ensure that she received medical aid" (Holmes, quoted in Grint and Woolgar 1997). Even death is not so certain. Leaving aside common uncertainties about causes of death and the timing of death, there are cross-cultural differences about death and what happens following it. No matter how unmalleable a technology might look, there are always situations, some of them highly hypothetical, in which the technology can take on unusual uses or interpretations. In addition, technologies are not as autonomous as they often appear. Some of the appearance of autonomy in a technology stems from of our lack of knowledge. As we gain knowledge of the historical paths of particular trajectories, we see more human roles in those paths (Hong 1998).

To accept that technologies do not have essences is to pull the rug out from under technological determinism. If they do nothing outside of the social and material contexts in which they are developed and used, technologies cannot be the real drivers of history. Rather these contexts are in the drivers' seats. This recognition is potentially useful in enabling political analyses, because particular technologies can be *used* to affect social relations (Hård 1993), such as labor relations (e.g. Noble 1984) and gender relations (e.g. Cockburn 1985; Cowan 1983; Oudshoorn 2003).

There should, then, be no debate about technological determinism. However, in practice nobody holds a determinism that is strict enough to be completely overturned by these arguments. Even the strictest of determinists admit that social forces play a variety of important roles in producing and shaping technology's effects. Such a "soft" determinism is an interpretive and heuristic stance that directs us to look first to technological change to understand economic change (Heilbroner 1994 [1967]). Choices are made in relation to material resources and opportunities. To the extent that we can see social choices as economic choices, technology will play a key role.

Anti-essentialists show us that even soft determinism must be understood within a social framework, in that the properties of technologies can determine social structures and events only once the social world has established what the properties of technologies are. Anti-essentialism directs us to look to the social world to understand technological change and its effects. This is perhaps most valuable for its constant reminder that things could be different.

On the other hand, we study technology because artifacts appear to do things, or at least are made to do things. Thus a limited determinism is right, given a particular set of actions within a particular social and material arrangement. "Our guilty secret in STS is that really we are all technological determinists. If we were not, we would have no object of analysis" (Wyatt 2007).

Bijker's theory of *sociotechnological ensembles* is an attempt to understand the thorough intertwining of the social and technical (Bijker 1995). Bijker draws heavily on work on technology as heterogeneous engineering, and on his earlier work with Pinch on SCOT. A key concept in the theory is that of the *technological frame*, the set of practices and the material and social infrastructure built up around an artifact or collection of similar artifacts – a bit like Kuhn's *paradigm*. Another, similar concept is the notion of a technological *script*, developed from within actor-network theory (Akrich 1992). As the frame is developed, it guides future actions. A technological frame, then, may reflect engineers' understandings of the key problems of the artifact, and the directions in which solutions should be sought. Technological frames also reproduce themselves, when enough physical infrastructure is built around and with them. The "system of automobility" (Urry 2004), for example, has been and continues to be very durable, given the enormous physical, political, and social infrastructure, on both local and global scales, based on cars made of steel that use petroleum.

A technological frame may also reflect understandings of the potential users of the artifact, and users' understanding of its functions. If a strong technological frame has developed, it will cramp interpretative flexibility. The concept is therefore useful in helping to understand how technologies and their development can appear deterministic, while only appearing so in particular contexts. For example, in recent decades research on contraception has concentrated on female contraception. This gender asymmetry and the assumptions behind it are constituted by negotiations, choices, and contingencies. As a result, new male forms of contraception depend on alternative sociotechnical ensembles. Ultimately, their success will depend on cultural work to reshape the features of gender that contribute to the asymmetry, and to which the available technologies contribute (Oudshoorn 2003).

Stretching the concept a little, we might see how technological frames can highlight some features of technologies and hide others. In the nineteenth century governments and banks went to extensive efforts to protect paper currency from forgery. Exquisite engravings, high-quality paper and ink, and consistent printing processes gave users of paper money more confidence that it was genuine. In principle those features of paper money could not serve to answer doubts about the solidity of paper currency – in comparison to coins containing precious metals – even if the banknote as a work of art had the effect of distracting from those doubts (Robertson 2005).

Architecture provides a good example of the ways in which technologies have effects and embody social structure. A building is a piece of technology, one that shapes the activities, interactions, and flows of people. For example, a study of the design and use of a university biotechnology building showed some of the negotiations incorporated into the building: it will have space for visiting industrial scientists, but not for undergraduate students; it will integrate molecular biology, genetics, and biochemistry, but separate molecular biology from microbiology (Gieryn 2002). Once these decisions are made, the (literally) concrete structure forms a concrete social structure. Yet, there is more flexibility than designers intended: the unused labs for visiting industrial scientists eventually become academics' labs; and some teaching of undergraduate students takes place in research labs. The building becomes reinterpreted.

Even if technologies do not have truly essential forms – properties independent of their interpretations, or functions independent of what they are made to do – essences can return, in muted form. Artifacts do nothing by themselves, though they can be said to have effects in particular circumstances. To the extent that we can specify the relevant features of their material and social contexts, we might say that technological artifacts have dispositions or affordances. Particular pieces of technology can be said to have definitive properties, though they change depending upon how and why they are used. Material reductionism, then, only makes sense in a given social context, just as social reductionism only makes sense in a given material context.

Does technology drive history, then? History could be almost nothing without it. As Bijker puts it, "purely social relations are to be found only in the imaginations of sociologists or among baboons." But equally, technology could be almost nothing without history. He continues, "and purely technical relations are to be found only in the wilder reaches of science fiction". (Bijker 1995).

Box 9.2   Were electric automobiles doomed to fail?

David Kirsch's history of the electric vehicle illustrates both the difficulty and power of deterministic thinking (Kirsch 2000). The standard history of the internal combustion automobile portrays the electric vehicle as doomed to failure. Compared with the gasoline-powered vehicle, the electric vehicle suffered from lack of power and range, and so could never be the all-purpose vehicle that consumers wanted. However, "technological superiority was ultimately located in the hearts and minds of engineers, consumers, and drivers, not programmed inexorably into the chemical bonds of refined petroleum" (Kirsch 2000: 4).

Until about 1915 electric cars and trucks could compete with gasoline-powered cars and trucks in a number of market niches. In many ways electric cars and trucks were successors to horse-drawn carriages: Gasoline-powered trucks were faster than electric trucks, but for the owners of delivery companies speed was more likely to damage goods, to damage the vehicles themselves, and in any case was effectively limited in cities. Because they were easy to restart, electric trucks were better suited to making deliveries than early gasoline-powered trucks; this was especially true given the horse-paced rhythm of existing delivery service, which demanded interaction between driver and customer. Electric taxis were fashionable, comfortable, and quiet, and for a time were successful in a number of American cities, so much so that in 1900 the Electric Vehicle Company was the largest manufacturer of automobiles in the United States.

As innovators, electric taxi services were burdened with early equipment. Some happened to suffer from poor management, and were hit by expensive strikes. They failed to participate in an integrated urban transit system that linked rail and road, to create a niche they could dominate and in which they could innovate. Meanwhile, Henry Ford's grand experiment in producing low-cost vehicles on assembly lines helped to spell the end of the electric vehicle. World War I created a huge demand for gasoline-powered vehicles, better suited to war conditions than were electric ones. Increasing suburbanization of US cities meant that electric cars and trucks were restricted to a smaller and smaller segment of the market. Of course, that suburbanization was helped along by the successes of gasoline, and thus the demands of consumers not only shaped, but were shaped by automobile technologies.

In 1900 the fate of the electric vehicle was not sealed. Does this failure of technological determinism mean that electric cars could be rehabilitated?

According to Kirsch, writing in the late 1990s, that seemed unlikely. Material and social contexts have been shaped around the internal combustion engine, and it seemed unlikely that electric cars could compete directly with gasoline cars in these new contexts. Yet, only a few years later, it appears that there may be niches for electric cars, created by governments' and individuals' commitment to reducing greenhouse gases.

10

Studying Laboratories

The Idea of the Laboratory Study

In the 1970s a number of researchers simultaneously began using a novel approach to the study of science and technology: They went into laboratories in order to directly observe practical and day-to-day scientific work. Most prominent among the new students of laboratories were Bruno Latour (Latour and Woolgar 1986 [1979]), Karin Knorr Cetina (1981), Harry Collins (1991 [1985]), Michael Lynch (1985), Sharon Traweek (1988), and Michael Zenzen and Sal Restivo (1982). At the same time, June Goodfield (1981) and Tracey Kidder (1981) published fictional ethnographies of science and technology that fit well with the new genre of the laboratory ethnography. Laboratories are exemplary sites for STS because experimental work is a central part of scientific activity, and experimental work is relatively visible (though Lynch (1991) points out that much work in the laboratory is not – researchers may sit at microscopes for hours at a time).

Many of the first students of laboratories used their observations to make philosophical arguments about the nature of scientific knowledge, but framed their results anthropologically. In their well-known book *Laboratory Life*, Latour and Woolgar announce their intention to treat the scientists being studied as an alien tribe:

> Since the turn of the century, scores of men and women have penetrated deep forests, lived in hostile climates, and weathered hostility, boredom, and disease in order to gather the remnants of so-called primitive societies. By contrast to the frequency of these anthropological excursions, relatively few attempts have been made to penetrate the intimacy of life among tribes which are much nearer at hand. This is perhaps surprising in view of the reception and importance attached to their product in modern civilised societies: we refer, of course, to tribes of scientists and to their production of science. (Latour and Woolgar 1986)

Treating scientists as alien gives analysts leave to see their practices as unfamiliar, and hence to ask questions of them that would not normally be asked. In line with the methodological tenets of the strong programme, an anthropological approach opens up the study of scientific practices and cultures.

The central question asked in the first round of laboratory studies was, simply, how are facts made? That is, how can, work in the laboratory give stability and strength to claims, so that they come to count as pieces of knowledge? Laboratory studies framed their answers to this as rejections of a taken-for-granted assumption that a laboratory method solved that problem. Instead, all of the early laboratory studies emphasized the *indexical* (Knorr Cetina 1981; Lynch 1985) nature of scientific reasoning and actions in the lab, reasoning and actions that are tied to circumstances and unpredictable in advance (Doing 2007). Moreover, decisions about claims are negotiated, a matter for different actors to decide through micro-sociological or political interactions. Conversations decide what it is at which the scientists are looking. Negotiations decide what can be written into manuscripts. Rhetorical maneuvers help shape what other scientists will accept. In general, scientific practices in the lab are similar to practices of familiar ordinary life outside the lab, and there is no strict demarcation between science as done in laboratories and non-science outside of laboratories. To put it polemically, nothing "scientific" is happening there (Knorr Cetina 1981).

The indexical nature of scientific reasoning means that there is no simple answer to the question of how facts are made. Nonetheless, two schema help to provide answers. The first is actor-network theory, which we have already seen: ANT claims that the laboratory is an important source of facts because the laboratory contains material tools for disciplining and manipulating nature, making it ready for the creation of general facts. The second schema emphasizes the development of unformalizable expertise, that can successfully discipline and manipulate nature, and that can generalize. This chapter emphasizes the second of the two. These two schema share much, but also appear to conflict (Chapter 12).

Learning to See

Ethnographers, as outsiders, did not see what laboratory scientists saw. The very things that scientists took as data were often difficult to recognize as distinct objects or clear readings. Unsurprisingly, experts have learned how to read their material in a way that novices cannot.

Pictures of such things as electrophoresis gels of DNA fragments or radioactively tagged proteins do not provide unproblematic data. They simultaneously contain too much information, and not enough. Experts may see the patterns that they are looking for, or see that those patterns are not present, perhaps after some puzzling and discussion, by making the pictures contain exactly the right amount of information (Zenzen and Restivo 1982; Charlesworth et al. 1989). At the same time, the fact that experts can see what is relevant suggests that they may miss other interesting features of their material (Hanson 1958; Kuhn 1970). They have to block out some features and emphasize others, turning the picture into a diagram in their mind's eye. Here is one description of the process:

> Kornbrekke's first problem was to determine what emulsion type he had after shaking. This sounds innocent enough, but it is fraught with difficulty. After shaking, most of the mixtures separate quite rapidly and one sees a colorless solution with a highly active boundary or interface between the two volumes of liquids. All the standard techniques for determining emulsion type or morphology apply only to *stable* emulsions. Kornbrekke had to teach himself to see what a short-lived emulsion "looks like". By "thinking about what you have" (in Kornbrekke's words) and hypothesizing relevant visual parameters, Kornbrekke had to make the data manifest themselves. (Zenzen and Restivo 1982)

Tinkering, Skills, and Tacit Knowledge

As Ian Hacking (1983) argues, laboratory work is not merely about representation, but about intervention: researchers are actively engaged in manipulating their materials. We have already seen something of the messiness of the laboratory. The messiness of observation finds its counterpart in the intervention side of laboratory life. Manipulations fail, apparatuses do not work, and materials do not behave. Hacking puts it bluntly, "experiments don't work" (Hacking 1983: 229). Laboratory research thus involves a tremendous amount of *tinkering* (Knorr Cetina 1981) or *bricolage* (Latour and Woolgar 1986) in order to make recalcitrant material do what it is supposed to. Prior rational planning only covers some of the situations faced in the laboratory, and needs to be supplemented by skilled indexical reasoning, tied to particular problems (Pickering 1995).

The idea of *tacit knowledge* is due to the chemist and philosopher Michael Polanyi (1958). Collins introduced the term to STS in a study of attempts by researchers to build a Transversely Excited Atmospheric Laser, or TEA-laser (Collins 1974, 1991). In the early 1970s, he interviewed researchers at six different British laboratories, all of which were trying to build versions of a TEA-laser, and interviewed scientists at five North American laboratories from which the British groups had received information. The main attraction of the TEA-laser, which had been developed in a Canadian defense laboratory, was the low cost and robustness of its gas chamber. The components could be easily purchased, and the design was thought to be particularly straightforward.

However, the transfer of knowledge about how to build the laser was difficult. Nobody "succeeded in building a laser by using only information found in published or other written sources. . . . no scientist succeeded where their informant was a 'middle man' who had not built a device himself. . . . [E]ven where the informant had built a successful device, and where the information flowed freely as far as could be seen, the learner would be unlikely to succeed without some extended period of contact with the informant" (Collins 1992: 55). The difficulty was not merely because of secrecy, as Collins observed cases in which there was no apparent secrecy, but in which knowledge did not travel easily. Rather, the transfer of knowledge about how to build the laser required more than simply a set of instructions; it required the passing on of a skill. As a result Collins compares two models of the transfer of knowledge: the "algorithmic" model assumes that a set of formal descriptions or instructions can suffice; the "enculturational" model assumes that socialization is necessary.

While some knowledge can be easily communicated in written (or some equivalent) form, some resists formalization. Some relevant knowledge about TEA-lasers could be isolated, written down and distributed, but some information could be communicated only through a socialization process. This is *tacit knowledge*. Tacit knowledge can be embodied knowledge, like that of how to ride a bicycle, how to grow a crystal, or how to clone an antibody; in some cases, bodily knowledge is a resource for discovery, as when crystallographers feel their way around real and imagined models of molecules (Myers 2008). Tacit knowledge can also be embedded in material and intellectual contexts, dependent upon particular material and intellectual arrangements for its success and even meaningfulness. Scientific knowledge has an important material and intellectual locality, a finding of almost every empirical study in STS. Collins has devoted a considerable amount of effort, and several books (Collins 1990; Collins and Kusch 1998) to arguing that some knowledge is essentially tacit, when it is irreducibly tied to social actions. Expertise can never be entirely formalized (see Box 10.1).

We can pick up another crucial aspect of scientific and technical skill from the TEA-laser study. In addition to interviewing scientists, Collins spent time in laboratories observing and helping with the construction of lasers,

Box 10.1 Limitations of artificial intelligence

Hubert Dreyfus's 1972 book, *What Computers Can't Do*, argues that there are domains in which artificial intelligence (AI) is impossible if the AI is based on mere facts collected from experts. This provocative claim flies in the face of the promises made by researchers on and promoters of AI since the invention of computers. Dreyfus's argument turns on a particular notion of tacit knowledge: there are subject matters that are impossible to completely formalize.

Even something as apparently structured as playing chess turns out to be based on much tacit knowledge. Chess players at the highest level do not apply theories of the game, and do not examine more possible moves than players directly below them in skill. They are better by recognizing situations as similar to ones they have seen before, and knowing what to think about for those situations. Chess programs have become very good, even defeating the very best at the game, but they operate on very different principles. They apply programmed strategic rules of thumb to enormous numbers of possible moves. The good chess player simply perceives patterns and similarities, and that ability has become part of his or her perceptual apparatus. Computers need to have definitions of patterns before they can be useful, and those are difficult to find, in chess and in most other domains of interest.

Collins also has a critique of AI based on thinking about tacit knowledge (e.g. Collins 1990). Expertise even about straightforward things is difficult to formalize. So, how does AI ever work? Collins claims that AI works precisely in those areas where people have chosen to discipline themselves to work in machine-like manners. Even then, computers give the appearance of working well because people easily, and sometimes unconsciously, correct their errors. Pocket calculators work well because people have forced arithmetic into rigid and formal modes. Yet it is not unusual for calculators to give answers like "6.9999996" in response to problems like "7 ÷ 11 × 11". Or, more importantly, to know your weight in kilograms if it is 165 pounds you type in "165 × 0.454" – and are given the answer "74.91." In normal human circumstances, though, the correct answer is "75" kilos, especially given that the starting point, 165 pounds, is marked as an approximation.

The developed version of this analysis starts from actions, not knowledge (Collins and Kusch 1998). "Polimorphic" actions – the spelling is right, drawing attention both to multiplicity and to their social derivation –

can be instantiated by different possible behaviors, only some of which are appropriate in any specific context. Writing a love letter, voting with one's conscience, and being playful are all polimorphic because the actions cannot be completed merely by copying previous instances of the behaviors that instantiate them: a love letter that is merely a copy of a previous love letter would be a failure or a deception. "Mimeomorphic" actions, on the other hand, are actions that can be performed by copying some limited repertoire of behaviors. While a computer-generated love letter does not, in general, count as a love letter, a computer-generated invoice counts all too well as an invoice – in fact, computer-generated invoices have come to look more appropriate than hand-written ones.

Thus we have two different analyses of the problem for AI. On the one, domains for which we have no theory are difficult to formalize. On the other, socially situated domains, in which human participants have not forced themselves to behave like machines, are impossible to formalize. These two claims are not contradictory, but they point in very different directions.

---

allowing him to see some of the difficulties firsthand. One of his subjects tried to make at least two of them for use in his lab. Both lasers were difficult to get in working order, even though the scientist appeared to have all of the needed technical knowledge and expertise. Almost every component and its placing that potentially caused trouble, and obvious frustration:

> I was then in a situation of – well, what shall I do with this damn laser? It's arcing and I can't [monitor the discharge pulse] so I thought well, let's take a trip to [the contact laboratory] and just make sure that simple characteristics like length of leads, glass tubes . . . look right. (Collins 1991: 61)

The first laser owed its working to considerable advice from the scientist's already-successful colleagues. But the second laser was also difficult to make, even though it was a near copy of the first! In complex contexts, then, technical skills are capricious. Judgments about what matters must be developed on the basis of considerable experience, but much science and technology operates in relatively novel contexts – dealing with new, contentious or poorly understood phenomena, or dealing with new, untested, or poorly understood techniques – and about which there is little experience. Graduate students discover this when they try to perform novel experiments, after an undergraduate career of duplicating already-performed experiments (Delamont and Atkinson 2001). So on the forefront of research, development, and education, many scientific and technical skills leave considerable room for uncertainty.

Creating Orderly Data

"Data" comes from the Latin for "givens," but much work has to be done to create typical data. While data are givens within science, the creation of data is a topic for investigation in STS. In a neurobiology laboratory, for example, photographs of sections are carefully marked to highlight the features that the researchers want to take as data (Lynch 1985). These features are not merely ones that would be invisible, indistinct, or unremarkable to an untrained eye, but only make sense in particular experimental and observational contexts. That is, the markings and enhancements of the slides bring to the fore features that the researchers, working in local research contexts, were looking for. So not only does it take expertise to read such slides – there are no mechanical rules for reading – but the expertise needs to be attuned to local circumstances. The careful highlighting and labeling of features on the photographs in some sense takes them out of their most local contexts and makes them available for inspection by the relevant expert community more generally.

There *are* generally applicable methodological rules, but work has to be done to make them applicable in particular contexts. Even within disciplines, such apparently straightforward things as measurement and observation, let alone more obviously complicated things as maintaining controls in experimentation, interpreting results, and creating models are local achievements (Jordan and Lynch 1992).

Since observation is not a straightforward process, how does data ever become stable enough to form the basis of arguments? How do researchers turn smudges into evidence? To see how, we might look at processes by which researchers in the laboratory decide on the nature of a piece of data, and by which they construct evidence for public consumption.

Many scientific and technical discussions are occasions on which judgments are clarified and organized. Here is a transcription of an exchange between collaborators, taken from Lynch's study of a neurobiology laboratory (Lynch 1985):

1   R: . . . In the context you've got . . . richly labelling in the (singulate)
2   (3.0)

3 (): Huh?
4 J: Mm mmm
5 R: I don't know whether there's an artifact
6   (1.0)
7 R: But [the
8 J: [Check the other sections.
9 R: They're the same.
10 J: They're the same?
11 R: Uhhmm noh,
12   (1 0)
13 R: Noh uh but uh but basically the same, yes.
14 J: There is something in- in th- in the singular cortex?
15 R: There is something
16 J: Well,
17   (1.0)
18 J: That would make sense.

Some conversational analysis turns up regularities and organization not just in the most formal version of exchanges, but in their pauses, interruptions, modulations of voice, and so on. Here, line (4) is taken as a weak "token of understanding" of (1), but given the long pause in (2) it is not taken as acceptance of the claim. Because J has held back acceptance, R raises the possibility of the labeling being an artifact. Following the demand in (8), that possibility is explored. (9) is an attempt to put it to rest, but is challenged. In (13), R asserts expertise in the matter, claiming that while the other slides are not *exactly* the same, they are *essentially* the same. Then, in (14) to (18), J lets the matter rest, accepting the original claim.

Claims are made and then adjusted in the context of the conversation. Participants are attuned to each other's subtle cues, and react to them, adjusting their claims to increase agreement, though sometimes to highlight disagreement. In this particular case agreement is reached without reference to particularly technical details; R's expert judgment that the other sections are the same for the purposes at hand is allowed to stand on its own.

While conversational analysis might seem to be overly attuned to minutiae, it shows how the work of conversation contributes to the establishment of facts. Especially when researchers are engaged in collaborative projects, and need to come to agreements, conversations display how researchers decide on the nature of data. Amann and Knorr Cetina (1990) investigate similar issues in a molecular biology laboratory. There, a key product was the electrophoresis gel, which when put on film shows indistinct light and dark bands that represent the lengths of different DNA fragments. These films are difficult to interpret even for experienced researchers, so there is often a period of consultation during which their nature is typically resolved, a process of "optical induction" (Amann and Knorr Cetina 1990):

Jo　if you shift <u>this</u> parallely with the others, right, like that, that way, that way, this is nonetheless not running on the same level as this

Ea　no, this isn't on the same level, granted. I am not saying (it is), but I say/this is/say/

Mi　but if these run on the same level, this greatly suggest, doesn't it, that this is the probe

Ea　sure this is the probe. But then I also know that I've got a transcript which runs all the way through

Jo　but this can be this/<u>this</u> one here. <u>That</u> is this band here. (Amann and Knorr Cetina 1990)

The biologists are deciding what they are looking at, drawing correspondences between different parts of the film, and between the film and the procedures that produced it. The result of such a conversation is that a complex array of light and dark bands is given an interpretation. It then becomes a defined piece of data to be drawn upon in later arguments.

Such conversations, which are often laborious and involve close interaction, cannot be part of published work. Other mechanisms must replace conversations in order to turn data into convincing evidence. Most straightforward, published images are very rarely the same as the images that are puzzled over in the laboratory. Published images of the electrophoresis gels discussed above were "carefully edited montages assembled from fragments of other images." These montages could then highlight the orders that had been achieved. Artifacts – unwanted by-products of laboratory manipulations – and other aspects of the original films that were considered unclear or irrelevant were trimmed from or de-emphasized in published photographs. The published fragments were juxtaposed to increase their meaningfulness, and "pointers" were added to push readers toward particular interpretations.

Images that are to serve as evidence in public contexts are (1) filtered, to show only a "limited range of visible qualities," (2) made more uniform, so that judgments of sameness are projected onto the images, (3) upgraded, so that borders are more sharply marked, and (4) defined, by being visually coded and labeled (Lynch 1990). Images become more like diagrams. Diagrams are not merely simplified representations, but are icons of what we might see as "mathematical" forms to which the researchers want to refer: "the theoretical domain of pure structures and universal laws which a Galilean science treats as the foundation of order in the sensory world" (Lynch 1990). While they may reflect the messy empirical domain, most untreated pictures of scientific objects are only poor indicators of the more clean and pure theoretical domain. Mathematized images superimpose order and structure on empirical observation; in this sense they are literary representations.

Similar issues apply to data more generally. Data or observations are the products of scientific work at a particular stage in the research: they are typically the products that feed directly into mathematical and statistical analysis. Data mathematize or digitize the objects of study (Roth and Bowen 1999). This is even though the work of collecting that data may be complexly indexical.

With all of this work, researchers eventually produce observations that can feed into the scientific analysis, observations that can form a basis of universal claims. Mention of the local work is generally not included in scientific publications, except insofar as it is describable in general methodological terms. Thus, the universal claims rest on observations produced by universal procedures, though the procedures are achieved locally.

The success of empirical work can be measured in the generality of the conclusions that are allowed to be drawn from it, or, in his more precise term, the *externality* (from the particulars of the study) of conclusions (Pinch 1985). "Splodges on the graph were observed" and "solar neutrinos were observed" are extraordinarily different statements of observations, though they might refer to the same event. The former is not likely to be challenged, but by itself has no importance. The latter stands a much higher chance of being challenged and may be of importance or interest. The goal of the researchers is to push their claims to as high a level of externality as they safely can, escaping the local idiosyncrasies of their work as much as possible.

Crystallization of Formal Accounts

In a widely read article, the biologist Sir Peter Medawar (1963) asked, "Is the scientific paper a fraud?" His argument was that empirical scientific articles are written as if they were narratives of events, but are also written to be arguments. Since sequences of events do not typically form good arguments, narratives are adjusted to guide the reader toward desired conclusions. Behind-the-scenes laboratory work like months of tinkering with experimental apparatuses, long discussions about the nature of data, false starts, and abandoned hypotheses are not included in formal accounts of an experiment. That is not even to mention such work as hiring and training research assistants, negotiating for money and laboratory space, responding to reviewers' comments, and so on.

Formal accounts provide a lens through which scientists and engineers see even their own work. In the TEA-laser study, after Collins's informants got their apparatuses working, they immediately attributed problems to human error. The machines with which they had been tinkering changed from being complex and unruly to being relatively simple and orderly. Before they were working, each component was a potential source of trouble; afterwards the machines had a straightforward design based on some simple principles.

The end of the research process involves a kind of *inversion* (Latour and Woolgar 1986). Through all of the early stages, there are doubts, disagreements, and above all work to define a fact. Almost all agency appears to belong to researchers. Once they have decided on it, though, they attribute reality and solidity to the fact. They deny their own agency, making the fact entirely responsible for its own establishment.

In the end, then, scientists and engineers assume that nature is orderly. Diagrams can stand in for pictures because diagrams represent a "mathematical" nature. Formal accounts can stand in for the weeks, months, or years experimenters spend getting their experiments to work because formal accounts represent the order of the experimental design. The work of the laboratory aspires to make manifest the order of nature, and thus it cancels itself out from the self-image of science.

Culture and Power

Most early laboratory ethnographies addressed philosophical questions about the production of knowledge. But others, and increasingly later ones, addressed questions about the cultures of laboratories *per se*. Sharon Traweek's *Beamtimes and Lifetimes* (1988) is the result of ethnographic studies of facilities in which high energy physics is done; in particular, she contrasts a US and a Japanese laboratory. Traweek studies the locations of power in these laboratories, and particularly how the physical spaces of the laboratories codes for, materializes, and enables that power. She investigates how the high energy physics culture tells "male tales" that create stereotypes of physicists as men (see also Keller 1985; Easlea 1986). Following traditional anthropological interests, Traweek explores physicists' different understandings of space and time, and in particular how different understandings of time – such as the time of physical events, reactor time, and trajectories of careers – interact. For Traweek, then, laboratory ethnographies are investigations of interesting cultural spaces of the modern world.

Laboratory sciences have different "epistemic cultures" (Knorr Cetina 1999). High-energy physics is oriented toward signs, as its objects are too small, fast, and short-lived to be observed in any ordinary sense, though they may be detected with elaborate equipment and simulated with large computer programs (Knorr Cetina 1999; Merz 1999). Knowledge production is done in large groups that function as organisms, though there is tension between these organisms and their individual scientist members. In contrast, molecular biology is a hands-on science performed in small-scale laboratories, and researchers' bodily skills are highly important both for manipulating the material and for observing it. Yet molecular biologists attempt to turn their small laboratories into factories that produce relatively large quantities of materials, which allow them to produce results (Knorr Cetina 1999).

Most of the first round of laboratory studies focused on local actions of and interactions among actors trying to create knowledge, and emphasized the contingency of local situations. But they largely left aside institutional features that shape what can be knowledge. However, industry patronage of some fields affects laboratory practice at very fine levels. A bacterial agent being studied in one plant pathology lab was evaluated in a particular agricultural context, subjected to tests against standard *chemical* fungicides to see its effectiveness against agriculturally important root rots (Kleinman 1998). Analogously, standardized technologies, which affect research intimately, are often embodiments of power relations: laboratories may not be in a position not to use them, or to use them in non-standard ways. Research is also shaped by assumptions about patents and about laboratories' relations to the universities in which they sit. Therefore, the institutional landscape should figure more in laboratory studies.

Nuclear weapons laboratories have been the subject of several valuable studies, including studies of both of the United States' laboratories (Gusterson 1996; Masco 2006; see also MacKenzie 1990). Because of the secrecy involved in nuclear weapons research, these studies are based on interviews, rather than participant-observation. At Lawrence Livermore National Laboratory, we see the scientists and engineers carefully disciplining themselves in their emotional understandings of their work (Gusterson 1996). Rather than being amoral bomb builders, they see themselves as contributing to détente, and concerned about the same issues as the protesters outside their gates. Yet there is a romance in their relations to the bombs they design, from the detailed precision of their workings to the enormous power of their effects. Ten years later, at Los Alamos National Laboratory, the weapons designers are engaged in technical work, relatively isolated – by the ban on testing – from any visceral experience of the weapons (Masco 2006). This is in contrast to activists' attention on the Laboratory, which is almost entirely focused on the destructive power of nuclear weapons. Meanwhile, nearby Hispanic and Pueblo groups are concerned with balances of interactions with the Laboratory, as it provides employment but also poses dangers to the local landscapes.

Extensions

The laboratory need not be neatly bounded by four walls, and laboratory studies need not describe only experimental work. Researchers have therefore extended the approach, problems, and style of early laboratory studies to other areas.

Theoretical physicists, for example, employ a variety of heuristics and tricks for expanding their theoretical objects into combinations of simpler ones. None of these take the form of rule-bound procedures, which makes experience and intuition about *how* to apply the heuristics valuable. Like the problems that experimenters face, the problems that theoretical physicists face cannot be solved by following neat recipes, but only by exploration, tinkering, sharing expertise, and eventually hitting upon the right answer (Merz and Knorr Cetina 1997; Gale and Pinnick 1997).

The insights gathered from laboratory studies also apply to field contexts. Agricultural scientists doing field trials are not farmers, and are therefore dependent upon the skills and cooperation of farm workers and farm owners. They have to devote considerable effort to assuring that cooperation (Henke 2000). A substantial part of the work of field research is work to turn the field into a laboratory; correspondences between the natural world and theoretical concepts can only be drawn if the natural world is physically transformed into representations (Latour 1999). Coming at a similar result from a different direction, a study of an oceanographic research vessel investigates how researchers from different disciplines work together to obtain samples, build representations of space, and coordinate multiple spaces. Correspondences have to be drawn between the bottom of the ocean floor, spaces created by the sonar that detects features of that ocean floor, spaces of the computer screen and spaces established by theoretical work. These correspondences require physical probes and markers that can be identified in different representations; they also constitute a complex coordinating activity by the researchers (Goodwin 1995).

The laboratory study taken to an artificial intelligence laboratory building a medical expert system can interestingly juxtapose AI researchers' understandings of knowledge as mere collections of facts with more anthropological understandings of knowledge (Forsythe 2001). Among other things, that juxtaposition reveals power relations within the lab between AI researchers and physicians whose knowledge they are trying to formalize. It is common among the AI researchers to "blame the user" for the failings of the expert systems they are trying to build, and for the failure of the medical establishment to use these systems. But the AI researchers' view, a feature of the cultural landscape of AI, of knowledge as something that can be extracted from physicians, results in an insulation of the systems from the real medical contexts in which they are meant to operate (Forsythe 2001).

Laboratory studies have even been applied to the social sciences. Economic modeling can be seen to be analogous to experimental work, involving skill and tacit knowledge, give-and-take with the models, and shaped by important cultural or technical assumptions (Breslau and Yonay 1999). Macroeconomic models can be interpreted flexibly, and are difficult to falsify (Evans 1997). In sociological surveys by telephone, skill plays a key role, even when the survey is ideally a purely formal instrument: good questioners deviate from their scripts in order to make their interactions with respondents more closely follow the formal model (Maynard and Shaeffer 2000). Even the apparently unlikely site of science policy has been approached via ethnographic methods, showing how science policy is constituted by a set of representational practices (Cambrosio, Limoges, and Pronovost 1990).

The laboratory study, then, has moved into new terrains. Original results are often seen again. However, especially with an interest in specific cultures, and in power, every new terrain is also a productive site for new insights.

11

Controversies

Opening Black Boxes Symmetrically

Science and technology appear to accumulate knowledge, piling fact upon fact in a progressive way. Although this may not be entirely accurate – Kuhn argued that science is less progressive than we take it to be, that there are losses in knowledge as well as gains – scientific and technological discussions seem to end with claims to solid knowledge more often than do discussions in most other domains. In this context STS appropriates the engineers' term *black box*, a term describing a predictable input–output device, something the inner workings of which need not be known for it to be used. Science and technology produce black boxes, or facts and artifacts that are taken for granted; in particular, their histories are usually seen as irrelevant after good facts and successful artifacts are established.

Once a fact or artifact has become black-boxed, it acquires an air of inevitability. It looks as though it is the best or only possible solution to its set of problems. However, this tends to obscure its history behind a teleological story. If we want to ask how the consensus developed that vitamin C has no power to cure cancer (Richards 1991), it hinders our study to say that that consensus developed because Vitamin C has no power to cure cancer. The truth has no causal power that draws scientific beliefs toward it. Instead, consensus develops out of persuasive arguments, social pressures, and the like. Similarly, if we want to ask how the bicycle developed as it did, it hinders our study to claim that bicycles with equal-sized wheels, pneumatic tires, particular geometries, etc., are uniquely efficient. The implausibility of that claim can be seen in the wide variety of meanings attached to bicycling and the wide variety of shapes of bicycles over the past 150 years (see Chapter 9; Pinch and Bijker 1987; Rosen 1993). Bicycles and bicycling have *interpretive flexibility*, with different meanings for different actors. There is no single standard of efficiency that draws artifacts toward it.

To disperse the air of inevitability, STS takes a symmetrical approach to studying debates. Rather than looking at facts and artifacts after they have been black-boxed, investigators pay particular attention to controversial stages in their histories. This might even involve studying genuinely open controversies, controversies that have yet to end. Researchers in STS have looked at genetically modified foods (Klintman 2002), parapsychology (Collins and Pinch 1982), the lethality of non-lethal weapons (Rappert 2001), the best strategies for artificial intelligence (Guice 1998), the definition of the moment of death (Brante and Hallberg 1991), and many other recent and current controversies. The model has even been extended to cover ethical controversies, such as that over embryo research (Mulkay 1994) or over genetic sampling of human populations (Reardon 2001), and voting controversies, such as that over the US Presidential election in 2000 (e.g. Miller 2004).

Reasonable Disagreements

There is a temptation to see the losing participants in controversies as unreasonable. However, we should take a more charitable attitude, especially toward scientific and other technical disputes. Almost all participants in disputes have reasons for their positions, and they, at least, see those reasons as good ones. A symmetrical approach attempts to show some of the force of those reasons, even the ones that eventually fail. STS attempts to recover rationality in controversies, where the rationality of one or another side is apt to be dismissed or forgotten.

For example, in the 1970s the geophysicist Thomas Gold proposed an "abiogenic" theory of the formation of hydrocarbons. Oil, he argued, is not a fossil fuel, but is a purely physical product. Gold's model of the formation of oil can be defended, and serious questions can be raised about petroleum geologists' standard models (Cole 1996). While Gold may be arguing for an unorthodox position, and perhaps even one that is difficult to maintain, it is not an irrational one.

Once again the Duhem–Quine thesis is relevant. It is always in principle possible to hold a theoretical position in the face of apparently contrary evidence. Though this looks like an epistemologist's claim with little applicability in practice, one can document cases in which scientists and engineers reasonably maintain positions – by plausible standards of reasonableness – even though other scientists and engineers may find conclusive evidence against those positions. The same can even be seen in mathematics, often thought to be a purely formal activity with no possibility for reasonable disagreement

Box 11.1   Cold fusion

In 1989, Stanley Pons and Martin Fleischmann, two chemists at the University of Utah, announced at a press conference that they had observed nuclear fusion in a table-top device. The announcement was stunning. It had previously been assumed that fusion, the process that produces the sun's energy and that is at the heart of a hydrogen bomb, could only take place in situations of very high energy, and could only be controlled with large and elaborate devices. Huge sums had been invested in conventional fusion research, largely in the hopes that controlled fusion reactions would lead to almost limitless energy. Pons and Fleischmann were apparently offering an inexpensive alternative.

Their fusion apparatus was simple and apparently easy to create. Many other groups rushed to repeat the experiments, initially working only from television and newspaper descriptions and images. Some of these groups immediately confirmed the initial results. Negative results appeared more slowly, but the tide turned against cold fusion in the months that followed. Meanwhile, theoretical arguments were created to show how cold fusion could have occurred, while other arguments were marshaled for the impossibility of cold fusion. Again, the tide turned against cold fusion. Disciplinary divisions were important, because fusion had been the preserve of physicists until that point, and many of them were quick to dismiss the two chemists who were cold fusion's primary discoverers. In a matter of months the skeptics had prevailed, and consensus was rapidly forming that cold fusion was the erroneous product of incompetent experimenters.

A number of researchers in STS have looked at the debate that followed Pons and Fleischmann's announcement, from a variety of perspectives. The case is interesting for how information and views of cold fusion circulated among researchers. The standard model of scientific communication emphasizes formal publications, but during this controversy researchers relied on every medium available to learn what was happening, including electronic bulletin boards, television and newspaper reports, and faxed preprints (Lewenstein 1995). The frenzy of communication helped to solidify views, as people learned about trends in the theoretical and experimental arguments. In the cold fusion controversy, replication of experiments was difficult, but it was quickly resolved without iron clad proofs or refutations (Collins and Pinch 1993; Gieryn 1992). The controversy interestingly shows how experimental work can be interpreted in different ways. Physicist Steven Jones at Utah's Brigham Young University, who had been in close contact with Pons and Fleischmann and had been working on similar experiments, presented his work much more modestly. Had Jones prevailed in his interpretation of the results, cold fusion might have survived under a different label, as an unexplained phenomenon worth exploring, though not one of earth-shaking importance (Collins and Pinch 1993). Finally, while consensus did quickly form against cold fusion, that consensus did not stop research (Simon 1999). The controversy, and as a result cold fusion science, was officially dead within a year, but many otherwise reputable researchers kept on doing experiments – hidden from mainstream science, being performed after normal working hours or in garages.

(Crowe 1988; MacKenzie 1999; Restivo 1990). Imre Lakatos's classic *Proofs and Refutations* (1976), for example, shows how the history of one mathematical conjecture is a history of legitimate disagreement (see Box 5.1 for Bloor's sociologizing of Lakatos's case).

At the same time, most minority views are eventually excluded from public debates. Extremely deviant views are marginalized, and more moderate deviants may be appeased and answered. (Though controversies are often framed in terms of neatly opposing sides, there are not usually just two sides in a controversy (Jasanoff 1996).) While challenges may be unanswerable in principle, they are in fact answered, assuaged, or avoided, allowing agreement and knowledge to be built up. Disagreements are regularly and routinely managed and contained.

Experimenters' Regress

Experiments are normally thought to provide decisive evidence for or against hypotheses. Because they are supposed to be repeatable, experiments look as though they provide something like solid foundations for scientific knowledge. The notion of *experimenters' regress* challenges the foundational distinctiveness of experimentation, and shows how there can be intractable controversies over experiments (Collins 1991). At the most abstract level it is another application of the Duhem–Quine thesis (Box 1.2).

At genuinely novel research fronts, experimenters do not know what their results will be, New results are, afer all, goals of experimentation (Rheinberger 1997). Experimental systems should be tools for producing differential responses, devices that reflect the different natures of different inputs, and in so doing have something to say. Medical researchers might have 50 patients take a drug that is believed to prevent heart disease, and 50 patients take a placebo. If the first group has significantly less heart disease than the second, this provides evidence that the drug does prevent heart disease, and thus the drug and the placebo are in this respect different. The experimental system is supposed to answer a specific set of questions about the drug. There are many other dimensions to experimental work, but the production of differential responses to different inputs bears the epistemic weight of experimentation.

Experimental systems created for doing real research (as opposed to teaching or demonstration) are generally relatively novel, because they are created to answer questions that have not yet been answered. Based on what we have seen, we could expect that it is often difficult to get an experimental system to work, to fine-tune it so that it responds in the right way. Because the TEA-laser was a piece of technology supposed to serve a particular purpose, it was relatively straightforward for the researchers to know when the laser was working: it burned through concrete (Chapter 10). But how do researchers know when an experimental system is working?

If experiments are supposed to answer genuinely open questions, then the experimenters, and the scientific community in general, cannot know what the answers are, and so do not have a simple yardstick to judge when the experimental system is working. The experimental system is working when it gives the right answer, but one knows the right answer only after becoming confident in the experimental system. Collins calls this the problem of experimenters' regress.

Experimenters' regress can become vivid when there is a dispute. In the early 1970s Collins studied attempts by physicists to measure gravitational waves. The theory of general relativity predicts that there should be gravitational waves, but the agreement among physicists at the time was that they would be too small to measure using available equipment. One researcher, Joseph Weber of the University of Maryland, developed a large antenna to catch gravitational waves, and started finding some that were many times larger than expected. This started a small flurry of attempts to replicate or disprove Weber's results. Some came out in his favor, and some came out against him. Who was right?

From Weber's point of view, people had attempted to replicate his results too quickly. He had spent years calibrating his antenna – one experimenter said that Weber "spends hours and hours of time per day per week per month, living with the apparatus" (Collins 1991). People who could not detect anything with their quickly developed gravitational wave detectors simply published their results. But were their detectors working well? On the other hand, Weber's results might have been artifacts of a device that was not measuring what he thought, or that was simply behaving erratically. His hours spent making his detector produce signals would be irrelevant if there were no waves to detect.

When should one experiment count as a replication of another? Each of the gravitational wave detectors in this controversy was different from each of the others. This is not merely because it is essentially impossible to create complex novel devices that must be considered exact copies of each other. Scientists rarely want to copy somebody else's work as exactly as possible. Even in the uncommon instances when they are trying to replicate somebody else's experiment, novelty is a goal: they want to refine the tools, try different tools and different arrangements, and apply particular skills and knowledge that they have. So there are no strict replications.

Even when one experiment is counted as replicating another by virtue of similarity, that may only serve to answer one set of worries. In principle, experiments counted as identical may also be thought to suffer from the same faults, and thus to produce the same poor results.

As the problem of foundationalism (Box 2.2) suggests, with enough work it is in principle possible to undercut the support for any claim. Experimenters' regress is a theoretical problem for all experiments, though only sometimes becomes a real or visible issue. Most experimental results are uncontroversial relative to the amount of work it would take to challenge them effectively, and if an experimental system is well established or well constructed, the results it produces will be difficult to dislodge.

Interests and Rhetoric

Controversy studies reveal the processes that lead to scientific knowledge and technological artifacts. In the midst of a controversy, participants often make claims about the stakes, strategies, weaknesses, and resources of their opponents. Therefore, researchers in STS have access to a wider array of information when they look at periods of active controversy than when they look at periods after controversies have been resolved.

What leads the central protagonists of a controversy to take their positions, particularly unorthodox positions? This question is only sometimes asked explicitly. Interestingly, not asking it amounts to assuming that the positions adopted by key participants are of intrinsic rather than instrumental value, that those participants adopt the perspectives they do simply because they find them attractive or plausible. Why, for example, did Thomas Gold take on well-entrenched beliefs and claim that petroleum is the result of purely geophysical processes, not biological and geophysical processes combined? Because of the recognition he would gain, especially if he was right? Because of the economic importance of his hypothesis? Because he enjoyed challenging orthodoxy?

If we ask the question explicitly, we may turn to either interest models or contextual explanations. For example, intellectual positions may cohere with social positions (e.g. MacKenzie 1978). Positions may also cohere with past investments in skills, resources, and claims: positions fit better or worse into overall programs of research. In a dispute over the origins, dates, and importance of the archaeological site Great Zimbabwe, various professional interests and established positions clashed with each other, mixed with familiar conflicts over land and race (Kuklick 1991). Looking at technological controversies, the interests at stake are more obvious: In one biotechnology patent dispute, for example, the sides were competing companies each of which had a strong financial stake in the outcome, stemming from investments in research and expertise (Cambrosio, Keating, and Mackenzie 1990).

What tools do actors employ to further their positions? Scientists and engineers need to convince people of their claims, and therefore rhetorical tools are central. In science in particular, some rhetoric is easily available for study, because a main line of communication is the published paper, an attempt to convince a particular audience of some fact or facts. In both science and technology, less formal lines of communication, such as face-to-face interactions, are more difficult to study, but they are no less rhetorical; thus informal communication can also be studied in terms of its persuasive power (Chapter 13).

An emphasis on persuasion is not to deny the importance of new pieces of evidence, for example performing new experiments to support a certain claim. However, viewed from the perspective of published papers, rhetoric always mediates such things as experiments and observations, standing between readers and the material world. That is, experiments and observations are presented to readers, so they are inputs to the rhetorical task. In addition, empirical studies are designed with persuasion in mind: a study that has no potential to convince audiences is a poor one. Since rhetoric is the main subject of a later chapter, the discussion here will be brief.

A central rhetorical task in a controversy is to convince audiences of the legitimacy of one's positions, and the illegitimacy of those of opponents.

To a large extent this means making one's own work appear more scientific, or more central to key traditions, than that of one's opponents. Thus one of the most important rhetorical resources is the idea of science itself. In the well-known dispute over plate tectonics, for example, a number of the disputants invoked maxims about the scientific method to appropriate the mantle of science. Some participants portrayed themselves as more concerned with fidelity to data and thus more empiricist; some portrayed themselves as making their claims more precisely falsifiable; and some took the risky strategy of allying themselves with a Kuhnian picture of science (Le Grand 1986).

Similarly, disciplines are important. In the cold fusion saga, for example, the fact that two of the main proponents of cold fusion were chemists made them easy targets for their physicist challengers (Box 11.1). The physicists could argue that fusion was a topic and had a tradition within the province of physics, and dismiss the radical statements of a few outsiders. Meanwhile, the chemists could point to physicists' interests in the large sums of money available for conventional fusion research. More subtly, authors can invoke traditions simply by citing important canonical figures and portraying their work as following established patterns.

Appeals to reputation can serve something of the same function. Someone with a reputation for being a brilliant and insightful theorist, a careful and meticulous experimenter, or who simply has a record of accepted results might be believed on those grounds. Having worked with a respected colleague, being at a large research institution, and having a large laboratory are also aspects of reputation that might be invoked in the course of a controversy, as might opposite claims.

A very different way that authors can legitimize and delegitimize is by invoking norms of scientific behavior. To question other researchers' open-mindedness, whether by showing their commitment to broad programs of research or in the extreme by suggesting that they have financial stakes in the outcomes of research, is to question their disinterestedness. Even humor and creating the appearance of "farce" can be used to isolate positions (Picart 1994). Norms of scientific behavior, such as the ones Merton identified (Chapter 3), can be mobilized in the service of particular ends.

More broadly, however, a scientific paper is designed to convince its audience. A well-constructed article is designed to convince readers through the data it brings to bear, the other articles cited, its lines of argument, the language it uses, and so on. Scientific writing, like most other writing, is constructed to have effects, and when it is carefully done all of its elements contribute to those effects. For example, an article may contain criticisms of assumptions, studies, experiments, or arguments made by opponents in the controversy. This criticism may be blunt or subtle, straightforward or technical – as issues become more controversial they tend to become more technical, but even technical criticisms are constructed to convince.

Technological Controversies

Technological controversies, if they can be isolated as such, follow similar patterns. Whether the issue is nuclear energy (Jasper 1992), the choice between different missile guidance systems (MacKenzie 1990), or the choice between electric and gas-powered automobiles (Box 9.2), technical controversies do not have straightforward technical solutions, at least not if those solutions exclude human factors. For example, in patent disputes, issues revolve around the question of whether a commercially valuable innovation is "novel" and "non-obvious" – whether it should count as a genuine invention or merely the use of an old invention. There is no merely technical answer to this question, in the sense of one that stands apart from issues about the nature and constitution of the expert community, from issues about standards of novelty, and from issues about the histories of their development. Technical issues can be seen to be simultaneously social, historical, economic, and philosophical (Cambrosio, Keating and MacKenzie 1990). There is no neat way to cut through the complexity to be left with a simple, definitive, and non-human solution.

The result of this complexity is that when people, whether they are engineers, investors, or consumers, face choices among technologies, there is no context-independent answer as to which is better. They face choices that must navigate among constellations of interests, goals, claims, images, and existing and potential artifacts. Choices become more complex when even the most narrowly technical characteristics of competitors may be in dispute, because tests are never definitive in and of themselves (Box 5.2). And because technological issues tend to impinge on public concerns more directly than do scientific issues, external actors can quickly become involved in technological controversies. Questions about particular technologies' capabilities and characteristics can become issues to be resolved by non-experts as well as experts. For example, whether non-lethal weapons, such as pepper spray or rubber bullets, are safe or dangerous becomes an issue decided – or left ambiguous – by the interventions of a wide variety of interested parties (Rappert 2001).

Box 11.2   The Challenger launch decision

Diane Vaughan's *Challenger Launch Decision* (Vaughan 1996) is a thorough examination of a single technological decision, albeit one with large consequences. The January 28, 1986 explosion of the Space Shuttle *Challenger* is a well-known event, in part because of the vivid image of its explosion, in part because of the inquiry that followed, and in part because it seemed to symbolize the vulnerability of the most sophisticated technological systems to small problems. The immediate cause of the accident was the failure of an O-ring that was supposed to provide crucial sealing between the joints of the Solid Rocket Motor – a point dramatized by physicist Richard Feynman in the subsequent inquiry (Gieryn and Figert 1990). NASA had asked the manufacturer of the Solid Rocket Motor, engineering firm Morton Thiokol-Wasatch, about the effects of cold weather during the launch, and the Thiokol engineers responded by pointing out the possible problem of O-ring failure. Their interpretation of data supporting their concern was questioned, however, and the launch went ahead.

Vaughan's study is an attempt to understand the culture at NASA, and how the decision to launch the *Challenger* was made, given the safety concerns. Ultimately, the decision was fully in accordance with NASA protocols, in conjunction with the data on O-ring failure, and a normalization of risk following previous O-ring failures. Blame appears to lie in the culture and organizational structures of NASA. They too explicitly accepted the risk involved in every decision, and were not prepared for the problems they faced in January 1986.

William Lynch and Ronald Kline (2000), applying Vaughan's work to engineering ethics, argue that simply introducing engineers to moral vocabularies and moral reasoning is unhelpful, because that approach typically leaves dissenting engineers with only an unattractive role to play, that of the whistleblower. Attention to ways in which procedures can make each step of an unacceptable path rational, and to ways in which risk is incrementally normalized, may be more valuable than attention to grand moral confrontations.

Vaughan's position is not itself uncontroversial, a fact that may demonstrate the resilience of controversies. Edward Tufte (1997), in an argument on the importance of good visual displays of information, argues that had the engineers created a better graph of the data on O-ring failure, it would have become clear that conditions of the launch were very dangerous ones, and the Shuttle would not have been launched. Many people find Tufte's new graphs striking enough to cut through Vaughan's complex arguments. At the same time, Tufte's graph has itself been criticized for subtly changing the information, and the result is that it is unclear whether such simple blame can be assigned. So the meta-controversy remains unresolved.

The Resolution of Controversies

Given the Duhem–Quine thesis, and the impossibility of unchallengeable foundations, how are disputes in science and technology resolved? How is *closure*, the ending of active debate, achieved? Here are five different types of contributions, most of which will be brought into play in any given dispute.

1   Detailed critiques of observations, experiments, and positions

The most straightforward attempts to discriminate among positions start from criticisms and questions about such things as the consistency and plausibility of positions, the solidity of experimental systems, and the appropriateness of experimental or observational procedures. To be taken seriously in the first place, an experiment must be designed so that there are few obvious unanswered questions that could affect how results are interpreted. Nonetheless, as the underdetermination thesis teaches us, it is impossible to answer all possible questions. Determined opponents can place any part of the system under the spotlight and challenge its role. Of course, challenges can be answered, too. As one of the subjects in Collins's gravitational wave study said, "In the end . . . you'll find that I can't pick it apart as carefully as I'd like" (Collins 1992: 88). The force of a challenge will depend upon the relative balance of its and the response's plausibility, given the work that participants are able to do.

2   New tests, and calibrations of instruments and procedures

New tests can help to resolve the issue. Again, the underdetermination thesis shows that no data can be definitive. However, some tests can be difficult to challenge, because they are expensive or are produced by positions central to the field. Even without new data, calibration can help decide among results. Especially when they are complex or delicate, the accuracy and precision of instruments can be challenged. They might be measuring something different from what they are designed to measure. Or, they might simply need adjustment. To demonstrate the (un)trustworthiness of an instrument, one can test it on known quantities. However, just as there can be no perfect replications of an experiment, there can be no perfect calibrations of an instrument. the circumstances in which it is tested and refined are never exactly the same as the circumstances in which it is used (Box 5.2). Whether the differences make a difference can become a secondary issue about instruments.

3   Isolating one position as more scientific or central – or as deviant

The actions and strategies above are designed to legitimize and delegitimize different pieces of research. Legitimization and delegitimization is a more general category for resolving scientific and technical conflicts. Thus the strategies described in the last section to make work appear less or more scientific can also help end discussions. It is particularly important to solidify agreement among members of the *core-set*, the densely connected group of researchers whose opinions count most (Collins 1991). Public consensus in the core set can officially end a controversy even when peripheral or external researchers maintain deviant views. If the networks are strong enough that everybody who matters understands the consensus, people holding minority positions may find themselves entirely ignored if they continue publishing their views (Collins 1999). These people can relent, accept their marginalized status, or become dissident scientists, adopting alternative strategies for promoting their views and alternative views of science and its politics (Delborne 2008).

4   Showing one position to be more useful

Distinct from issues of validity and legitimacy are issues of pragmatics. Sometimes one idea will become dominant because many researchers can see how to use it, how to build on it, regardless of its validity. The triumph of Gauge Theory, the dominant physical model of particles, depended on the fact that many physicists already had the mathematical skills and knowledge needed to investigate the model, and therefore found it congenial (Pickering 1984).

5   *Ignoring deviant viewpoints and data*

A position that contradicts or runs against the grain of established scientific understandings is less likely to be accepted than one that agrees with expectations. If a position contradicts quite central beliefs then its proponents usually have to do substantial amounts of work before anybody will treat it seriously enough to bother engaging it. In some cases disputes about deviant ideas or results never arise, or are quickly forgotten, while most people in the field go about their own business. The potential dispute does not become an issue for the community as a whole, because rather than being refuted some results are simply ignored – a point made already by Kuhn (1970).

How to Understand Controversy Studies

Because STS looks at science and technology from the outside, applying frames of analysis not indigenous to science and technology, work in STS is sometimes viewed with suspicion. In particular, some scientists and engineers think that the field is hostile toward science and engineering, or that some of its concepts are attacks on the legitimacy of scientific and technical research. Experimenters' regress is a case in point. Allan Franklin, a physicist and philosopher, reads Collins's work on experimenters' regress as suggesting that experimental results are not to be trusted, and that the resolution of conflicting experimental results do not involve rational discussions (Franklin 1997). Looking at published reports, he reanalyzes the gravitational wave case, and argues that the evidence strongly supported Weber's opponents. In this he is undoubtedly right: most of the scientists involved took the matter as firmly settled, which means that they took the evidence as strongly against high fluxes in gravitational waves. If we look at the episode from the point of view of a participating scientist trying to figure out what to believe, we should expect to find ourselves evaluating much of the same evidence that was in fact used to settle the debate. The point of most work in STS, though, is not to simply check or challenge scientific and technical knowledge, but to understand its sources and meanings.

Readings like Franklin's are understandable, however. Symmetrical presentations of controversies are intended to show that disagreements can be legitimate, that there is a case for the heterodoxy. Because they are often intended as arguments against rigid and simplistic views of the scientific method, controversy studies are intended to show that there is no decisive evidence for or against scientific claims, that there is no agreed-upon formula that inexorably leads to one answer over the other. In addition, the results of controversy studies are not necessarily flattering to science and engineering, when, for example, they highlight the unruly processes of arguing. Controversy studies, then, might be seen as challenges to the science itself.

Researchers in STS are typically far too well versed with the details of their cases to suggest that closures of debates do not involve careful reasoning. They do not necessarily want to paint unflattering portraits, and they certainly do not hold that one should not believe scientific results or use technologies. Controversy studies, at least within standard STS frameworks, are aids to understanding the nature of the closure of debates. That is, symmetrical presentations of controversies highlight the resolution of debates as local and practical achievements that need to be understood in terms of the local culture. Why did a debate end which could, in principle, have continued indefinitely? If there were decisive interventions, what made them decisive?

When controversy studies work well, then, they show how evidence is tied to its local culture and contexts. By itself, some piece of data has no meaning. Data is only given meaning – as evidence – by the people who make use of it. Studies of scientific controversies show how people can give meaning to information and how they sometimes convince members of a community to agree with that meaning. They show how knowledge is built by a process of bootstrapping, but not that knowledge is groundless.

Captives of Controversies: The Politics of STS

While controversy studies may be symmetrical, they are rarely neutral. By showing the mechanisms of closure, controversy studies tend to be viewed as supporting the less orthodox positions. Therefore, the results of the studies can themselves become part of the controversy, picked up by one or more sides, probably by the underdogs. Controversy studies, and the researchers who perform them, run the risk of being "captured" by participants (Scott et al. 1990). Especially since uses of the study are somewhat unpredictable, this can be worrisome for STS researchers.

How likely is it that controversy studies will be appropriated by one or another side in the controversy? Evelleen Richards (1996) reports that her earlier study of the controversy over vitamin C and cancer is viewed as supporting alternative medicine in its struggle to have the positive effects of vitamin C recognized. Organizations that support alternative medicine sell copies of her articles, and articles on alternative medicine cite her as exposing "the corruption at the heart of the cancer hierarchy in America," among other things (Richards 1996: 344). Nonetheless, she sees herself as articulating a position distinct from those of both alternative medicine and the medical establishment. A study of a variety of roughly "New Age" movements (Hess 1997) was similarly picked up by those movements, and criticized by their opponents.

However, there is some unpredictability in the "capture" phenomenon, an unpredictability that could just as easily give comfort as discomfort to the student of controversies (Collins 1996). In general, published work is available for use, and authors have little control over that use. The most unpredictable aspect of "capture" may be capture itself, because it is unclear how many STS controversy studies have affected their controversies. And there are many different levels of "capture," and some may be more or less irrelevant; it may not really matter within a scientific controversy that one or another side makes use of an STS analysis.

Nonetheless, the recurring possibility of "capture" makes the non-neutrality of STS's work clear. Perhaps, the field should not pretend neutrality, and researchers should make their commitments explicit (Scott et al. 1990). This would both allow their work to have more practical value, taking it out of the ivory tower, and allow the researchers to have more control over the uses to which their work is put. Since among the many roots of STS are some activist ones, this recommendation is attractive to some participants in the field.

There are different ways of making commitments explicit. The strongest version involves becoming an active participant in the controversy. Brian Martin (1996) reports on an "experiment" in which he helped to publicize an unorthodox theory on the origins of AIDS, the theory that it was transmitted via impure polio vaccines. His participation allowed Martin access to documents to which he would not otherwise have had access, although it also closed off his access to some people on the other side of the controversy. Although this experiment looks like an extreme case, it may not be: becoming a participant may be particularly easy when policy concerns are visibly intertwined with technical issues, as in many environmental controversies, medical controversies, and controversies over technologies in the public eye.

A more modest form of commitment involves simply making one's position transparent. Donna Haraway's (1988) metaphor of "situated knowledges" may be helpful here. Haraway argues that it is possible to simultaneously strive for objectivity and to recognize one's concrete place in the world. This "embodied objectivity" can produce partial knowledge, though knowledge that is in every way responsive to the real world. Accepting Haraway's analysis may mean that researchers can acknowledge the perspective from which they write and act, without relinquishing ideals of objectivity (Hess 1997).

Such a perspective may be as simple as a position from within the discipline. The perspective that a researcher in STS brings to bear on a controversy is unlikely to mirror that of any of the participants – even a symmetrical approach to the social constitution of scientific and technological knowledge should be distinct from the approaches of participants. Similarly, a properly analytical perspective may not be neutral between parties, but should be non-aligned. Arguably, a sophisticated analytic perspective from within STS is the only form that the commitment need take, providing a basis for sophisticated political positions.

12

Standardization and Objectivity

Getting Research Done

Research in the natural sciences tends to be more expensive than that in the humanities and social sciences, because tools and materials are more expensive. Researchers in the natural sciences tend to rely on each other, to be more *mutually dependent* (Fuchs 1992). If, in addition, tasks are relatively routine – if there is low *task uncertainty* (Fuchs 1992) – then there is considerable pressure to produce facts, as opposed to interpretations. Scientists will be interested in and potentially able to use each other's results, and therefore those results are made easily isolable and transportable. The natural sciences therefore often participate in a fact-producing cognitive style, typically standardizing and solidifying results.

New and innovative research areas are by definition less routine. They therefore have a higher task uncertainty. These areas should probably have a less authoritative style, and have relatively open boundaries. Disciplines with low material demands or well-distributed resources should have less mutual dependence, and consequently should have less social pressure to solidify results into factual nuggets; authority can be fragmented, and built on relatively local grounds, since there is less need to share resources. Thus disciplines with neither well-established tasks nor high levels of mutual dependence should have less stability, and hence more likelihood of continuing disagreement.

Communities that practice more standardized research are likely to be more coherent and successful. Having discovered how to create feasible research projects, they can both create and pursue them.

A similar analysis can be found in Joan Fujimura's (1988) study of cancer research (Box 12.1), which shows that a need to create stable work helped to shape cancer research. In the 1980s the molecular biological approach

Box 12.1   The molecular biological bandwagon in cancer research

According to Fujimura (1988), in the 1980s fundamental cancer research became reorganized around a standardized package of theory and technologies. The theory in question was the oncogene theory, which maintains that the normal genes that contribute to the necessary process of cell division can turn into oncogenes that create cancerous cells dividing out of control. This theory offers an explanation of widely different types of cancer, potentially unifying cancer research around a single step in the multiple pathways to cancer. The technologies of the package were recombinant DNA technologies, which have been becoming steadily more powerful since the middle 1970s.

In the early 1980s recombinant DNA technologies became developed to the extent that researchers could be sure of success in sequencing genes. Students could learn to use the technologies relatively easily, and could be set tasks that they could be relatively certain of completing within a fixed window of time. For both graduate students and senior researchers this is quite important: graduate students need to know that their research projects will not continue indefinitely, and more senior researchers want to know that they will have results before their grants run out.

Cancer researchers found themselves faced with a standardized package that promised avenues for successful research, avenues for funding, and a possibility of a breakthrough. As more and more cancer researchers adopted molecular biological techniques, and molecular biologists engaged in cancer research, they created a snowball by making one approach the norm, and marginalizing others. The rational researcher joined the bandwagon and contributed to it. It is worth quoting one of Fujimura's interview sources at length, for this researcher summarizes the process neatly:

> How do waves get started and why do they occur? I think the reason for that has to do with the way science is funded. And the way young scientists are rewarded . . . [A] youngster graduates and . . . gets a job in academia, for instance. As an assistant professor, he's given all the scut work to do in the department. . . . In addition, he must apply for a grant. And to apply for a grant and get a grant . . . nowadays, you have to first show that you are competent to do the [research], in other words, some preliminary data. So he has to start his project on a shoestring. It has to be something he can do quickly, get data fast, and be able to use that data to support a grant application . . . so that

he can be advanced and maintain his job. Therefore, he doesn't go to the fundamental problems that are very difficult. . . . So he goes to the bandwagon, and takes one little piece of that and adds to a well-plowed field. That means that his science is more superficial than it should be. (quoted in Fujimura 1988)

to fundamental cancer research overtook all of the others. There was a molecular biological "bandwagon" in cancer research (Fujimura 1988). This bandwagon was facilitated by the development of a package consisting of a theoretical model for explaining cancer and a set of standardized technologies for exploring the theory. A standardized technology is one that has become a black box (Latour 1987), or for which the contexts of successful use are flexible (Collins and Kusch 1998). Standardized technologies are shaped so that the skills required to use them are very generally held or quickly learned.

Scientific research, especially laboratory research, takes funding: equipment, samples, and materials have to be bought, research assistants have to be paid. But funding agencies are much more likely to support scientists who have established track records of successful research. The *cycle of credibility* (Latour and Woolgar 1986) is the cycle that allows scientists to build careers and continue doing research. Continuing research is central to the identity of most scientists, more so than, for example, earning money – even with the rise of a patent culture oriented around commercialization of results, the reward is often funding to do more research (Packer and Webster 1996).

To gain funding scientists typically write grant applications to public and private funding agencies. The success of grant applications depends upon evaluations of the likelihood of concrete results coming out of the research project. That in turn depends upon issues of credibility and direct evaluations of the *doability* of the project itself (Fujimura 1988). Thus scientists have a large stake in finding doable problems, both to gain immediate funding and to build up their base of credibility for the future. A standardized package is attractive because it helps to routinize research, research that is intrinsically uncertain. The pressures of continuing research, then, can easily push towards a type of mechanical solidarity in which researchers in a field do their work in standardized ways. One of the effects of this standardization is that members of a field can see each others' work as meeting standards of objectivity.

Absolute Objectivity

Although "objectivity" is a slippery term, most senses of objectivity relevant to the study of science and technology can be grouped in two clusters: *absolute* and *formal* objectivity. In the context of science, absolute objectivity is the ideal of perfect knowledge of some object, knowledge that is true regardless of perspective. Absolute objectivity is the "view from nowhere" (Nagel 1989). Formal or "mechanical" objectivity, on the other hand, is the ideal of perfectly formal procedures for performing tasks. In this sense the ideally objective scientific researcher would be machinelike in his or her following of rules.

Given the interpretive nature of scientific knowledge, it is difficult to make sense of the notion of absolute objectivity as anything other than a vague regulatory ideal. Nonetheless, one can trace the history of that ideal. For example, the drive for certainty or objectivity has some origins in an attempt to disconnect philosophy and science from political and economic uncertainty. It is plausible that the assassination of Henry of Navarre was highly symbolic for the young René Descartes, whose philosophical work is associated with a new emphasis on objectivity (Toulmin 1990). Henry of Navarre stood for a humanist tolerance of divergent views, of which religious views were key to the fate of Europe; after his assassination there remained little possibility in Europe for political moderation. In place of the failed humanist tolerance, Descartes attempted to install austerity and distance on the part of the knower, and stances that did not depend in any way on the values and views of the individual subject.

One interpretation of the ideal of absolute objectivity is in terms of value-neutrality. This latter concept has changed with changing historical and political contexts: it has stood variously for a separation of theory and practice, for the exclusion of ethical concerns from science, and for the disenchantment of nature (Proctor 1991). For nineteenth- and twentieth-century German academics, value-neutrality became an ideal in the social sciences because it served to insulate and protect science and the university from external controversies. This stance was articulated in opposition to particular social movements and used to combat those social movements: values should be separate from science because specific values should be separate.

There is no obvious way of making sense of absolute objectivity in practice. Quite literally, there is no view from nowhere in science, technology, and medicine (Livingstone 2003; Mol 2002). However, the judgments that experts make can stand in for absolute objectivity. Expertise is the ability to make (what are perceived as) good judgments, so expertise presupposes a grasping of the truth, if not distanced passionless knowledge. The truth that expertise is supposed to provide is objective truth, containing in itself no contribution from the expert. We will return to this notion of expertise as a stand-in below.

Formal Objectivity

In contrast to absolute objectivity, we can observe approximations of formal objectivity. We might see an analogue in the history of units of measurement and attempts at their standardization, in conflicts between local and global measures (Kula 1986). Before the nineteenth century, many European units of measurement, of different kinds of commodities and objects, were both local in definition and depended upon the qualities of what was being measured. A good example comes from Bourges in late-eighteenth-century France:

> A *seterée* of land is the only measure known in this canton. It is larger or smaller depending on the quality of the soil; it thus signifies the area of land to be sown by one *setier* of seed. A *seterée* of land in a fertile district counts as approximately one *arpent* of one hundred *perches*, each of them consisting of 22 feet; in sandy stretches, or on other poor soils, one *arpent* consists of no less than six *boisselées* . . . (quoted in Kula 1986)

Measurement required both local knowledge, and expertise in how to measure. Standardization movements were attempts to eliminate that necessary expertise. From a local point of view, though, standardized units were not always preferable. Not only were they unfamiliar, they changed the nature of measurement. For a farmer, the local *seterée* described above could be a more useful measure than a standardized *arpent*, and certainly more useful than the unfamiliar *hectare*. From a less local perspective, such as the perspective of a national government interested in collecting taxes, the *seterée* is a hindrance to general knowledge and efficient governance.

Something similar takes place in science and technology. Standardization of tools, metrics, units, and frameworks creates a kind of universality and eliminates subjectivity (O'Connell 1993; Gillispie 1960). There is a sense, then, in which all histories of the progress of scientific knowledge are histories of formal objectivity.

This even includes the history of the valuing of isolated facts. Some European intellectual circles of the seventeenth century attempted to curtail theoretical arguments by creating a new emphasis on and interest in isolated empirical facts, particularly experimental facts (Daston 1991). Although the concept of formal scientific objectivity as such is a later invention, agreements about isolated facts provide surrogates for objectivity, since these facts were not as open to question as were theories. The term "objectivity" does not come to take its current meanings and status until the nineteenth century, though, when expanding communities of inquirers made the elimination of idiosyncrasy more important. For example, among anatomists and pathologists, the search for objective representation was a moral, and not just a technical, issue, and was connected to a new image of the scientist as a paradigm of self-restraint (Daston and Galison 1992). The choice of specimens for atlases of the human body became a point of crisis, provoking discussions over the merits of different types and ranges of specimens that should be presented. Interest in eliminating choice also led to increasing use of tools for mechanical representation, such as the photography, which could be trusted for their impartiality, if not their accuracy.

Formal objectivity and informal expertise can easily come into conflict: objectivity is a form of regulation that limits the discretion of knowledgeable experts. Whereas much of recent STS has pointed to the indispensability of expertise and informal communities, rules and formal procedures are also important and have merited attention. Particularly in public arenas, objectivity, in the form of ever more precise rules and procedures to be followed, is a response to weakness and distrust (Porter 1995). During the Depression, the American Security and Exchange Commission changed the practices of corporate bookkeeping (Porter 1992b). The Commission insisted that assets be valued at their original cost, rather than their replacement cost. To accountants the latter is a better measure of the value of an asset, but is prone to manipulation. The Security and Exchange Commission, though, was interested in maintaining public confidence in accounting practices, not in realistic appraisal. In the right circumstances, then, an apparatus that gives a standardized response to a standardized question can be preferable over a person who gives more nuanced responses to a wider variety of questions, even if standardization gets in the way of truth. Moreover, formal objectivity in the sense of rules for scientists and engineers to follow can often be a response to epistemic challenges. When outsiders mount strong challenges to the authority or neutrality of scientists and engineers, the response is often to establish formal procedures that unify by removing discretion. Thus, "democracy promotes objectivity" (Porter 1992a).

Sometimes, though, challenges can prompt a reduction in standardization, when the rhetorical situation is right. For most of the twentieth century, fingerprinting was an unproblematic way of positively identifying individuals. The acceptability of fingerprint evidence in courts is largely due to standardization (in different ways and to different degrees, depending on jurisdictions) and an effacement of the place of skill. Thus, forensic fingerprint experts have been able to testify in courts that a single print left at a crime scene definitely matched one taken from a suspect. Court challenges, however have recently pushed fingerprinters to abandon standardization as the grounds of their authority. Challenges have, for now, gone hand in hand with a new acknowledgement of skill in the fingerprinting community, and an acceptance that experts can reasonably disagree (Cole 2002).

In engineering, standardization of artifacts is a related issue, though the challenges may be more obviously economic and political. A world of interchangeable parts is an efficient world. New parts can be simply bought, rather than commissioned from skilled artisans. Important to the history of standardized parts were the efforts of French military engineers of the late eighteenth century to buy standardized weapons (Alder 1998). To work, cannons and cannonballs had to match each other, and so had to be of quite precisely standard sizes, even though they were made by different artisans in different places. To enforce precision, the engineers developed the notion of tolerance, upper limits on the deviation from specified shapes and sizes. And to enforce tolerances, they developed ingenious gauges. The result was a set of mechanical judges of performance, which allowed inspectors to decide, and artisans little opportunity to contest, which artifacts were the same as the standards.

Formalized knowledge tends to command higher value for itself than tacit or artisanal knowledge. The "golden hands" that might be crucial to scientific success are recognized within a laboratory, but do not play a role in publications coming out of that laboratory. Tacit knowledge, then, is valued locally, but not publicly. The same is the case for craft knowledge. In the eighteenth century French and English engineers "rediscovered" Roman techniques for making cement that would harden under water, a recognized technical achievement. However, their rediscovery was also a formalization of artisanal knowledge that had been occasionally applied to canals and other structures. Knowledge about and practices involving Roman cement had been lost to science, but not to masons and some military engineers (Mukerji 2006). Pigeon-fanciers in the eighteenth and nineteenth centuries had considerable knowledge, much of it tacit, about selective breeding for very particular characteristics. Their knowledge, views, and writings were, however, ignored by naturalists before Charles Darwin. Darwin, as part of his unusual effort to gather systematic information about analogies to natural selection, went to considerable effort to understand the practices and culture of the pigeon-fanciers, and even acquired and bred pigeons himself. This great naturalist's *The Variation of Animals and Plants Under Domestication* brought pigeon-fanciers' craft knowledge into contact with science, in a way that both used it and kept it very separate (Secord 1981).

What About Interpretive Flexibility?

We have seen that when groups of experts face strong challenges they can respond by creating formal rules for their behavior. Even in the absence of challenges, science and technology gain power from the solidarity and efficiency that formal objectivity creates. Actor-network theory makes a related claim: science and technology gain power from the translation of forces from context to context, translations that can only be achieved in a rule-bound way. Intuitively, this appears indisputable. As was argued in the discussion of ANT (Chapter 8), in general the working of scientific and technical theories appears miraculous unless it can be *systematically* traced back to local interactions.

However, in practice, formal rules have to be interpreted. Wittgenstein's problem of rule following shows that no statement of a rule can determine future actions, that rules do not contain the rules for their own application. This problem is not just a theoretical one. Norms can be flexibly interpreted, and so do not by themselves constrain actions (Chapter 3). Mere instructions cannot typically replace the expertise that scientific and technical work demands. Expert judgment is ineliminable in science and technology, and much else.

Thus there are difficulties for the formal model. Formal rules cannot by themselves determine behavior or eliminate individual judgment. Rules do not, then, constrain actions as much as the proponents of objectivity might want to claim; instead, rules create new fodder for creative interpretations. In many complex situations formal rules cannot successfully replace expertise. There are always cases that, because of ambiguity, instability, or novelty, do not fit neatly into established categories. Experts may be able to recognize these cases and respond accordingly. Therefore, formalization is not always desirable. This is especially true in science and technology, because by definition research operates at the edges of knowledge.

The humanist model that makes informal expertise central to science and technology (Chapter 10) faces an equally substantial problem, precisely the problem that the formal model of objectivity claims to solve. The humanist model starts from the observation that expertise can be communicated via social interaction. Either social interaction reduces to a transfer of discrete pieces of knowledge, contrary to the humanists' arguments, or it contains a mysterious force. The first of these options is plausible. Even if social interaction is the best shortcut for communicating many forms of expertise – which it undoubtedly is – what could possibly pass between people that could not also be represented as discrete pieces of knowledge? Social interaction can efficiently orient attention in the right way, and clear away questions and misconceptions, but because of discrete information passing between people. Any other option appears to invoke the fundamentally mysterious force. Yet if science crucially depends upon a mysterious force, then the humanist model will fail to explain the success of science. If science works only because scientists are experts on the natural world, then we need to understand the sources of that expertise and its communication. These sources of expertise are precisely what the humanist model does not offer.

We are left needing to reconcile the humanist position that insists on the necessity of expert judgment with positions that insist on the power of formal rules. Both claim to describe essential features of science and technology, and both of those claims make sense and are backed by evidence. Clearly, we need to be looking at what is added to formal rules to make social activities orderly, what is added to allow people to coordinate actions, what is added to allow for communication.

A Tentative Solution

Wittgenstein showed that rules do not have any quasi-causal power, because any new action can be construed as being compatible with a given statement of a rule. But if this is true, why are social activities orderly? How do people coordinate their actions? Ethnomethodology approaches these questions by closely examining local interactions, to see how people produce and interpret social order (Lynch 1993). As its name suggests, ethnomethodology is an approach to studying methods particular to cultures or contexts. Commonsense knowledge is the central object of study, because actors' commonsense knowledge produces social structure. Beyond that, ethnomethodology is a notoriously nuanced approach; what follows is an interpretation of that approach for the purposes of addressing the conflict here.

At the level of local interactions, people do considerable work to create order. But there can be a special place for rules in this ordering. For example, telephone interviews in social scientific survey research are supposed to follow scripts; however, the people being called often do not want to answer the survey, and refuse to answer, or evade the interviewer. A skilled

Box 12.2   A sociology of the formal

In research on the computerization of record keeping in a hospital, Marc Berg (1997) begins work on a "sociology of the formal." A computer at each hospital bed is fed information from various monitors, and nurses and doctors are expected to make regular entries. The result is an active database, which calculates patients' fluid balances and plots their status. The formal tool shapes the work done by nurses and doctors. The fields to be filled in pattern the nurses' and doctors' actions by demanding information. However, the computerized record is descended from paper records, and the paper records themselves reflect prior practices – though each incorporates changes based on new compromises of ideals, goals, and limitations. As a result, each stage in the development of the record changes practices only slightly. There is a co-evolution of the work done in the hospital and its representation in the formal records. The computer does not suddenly create a new, Taylorized regime. Rather, the work done by nurses and doctors is shifted by the presence of the computer. Similarly, the computer does not suddenly create a new, abstracted patient. Rather, the view of the patient is closely descended from the view already created by well-established practices.

But people work with computers, rather than simply have their work shaped by them. The nurses and doctors routinely correct the information on the computer screen, make fictional entries to allow their work to proceed, and otherwise creatively evade the limitations of the system.

The nurses and doctors (and no doubt the hospital administrators, as well) also take advantage of the new power of the formal tools. "The tool selects, deletes, summarizes; it adds, subtracts, multiplies. These simple subtasks have become so consequential because they interlock with a historically evolved, increasingly elaborate network of other subtasks" (Berg 1997: 427). Formal tools afford new opportunities for creating order, and opportunities for new orders.

interviewer is constantly deviating from the script to bring the conversation back to it (Maynard and Schaeffer 2000). Following the script rigidly stands too large a chance of allowing the subject to not answer the survey, whereas deviations may help convince the subject to keep talking. At issue here is a goal, to get through the survey. This goal might be seen as a rule: complete the script, in the sense of learning subjects' reliable responses, *so that they can be used for objective research*. Such a rule, though, does not dictate what the speakers say and do. Instead, its status as a goal makes it a product of what they say and do. Completing the script depends on small deviations from the script in order to allow the conversation to continue. People attempt to align behavior with the rule, but are not governed by it. The result is a future-oriented object, the completed survey, which can be used by researchers to create more knowledge (see Miettinen 1998).

Many rules, then, do not dictate behavior, but instead serve as goals or ideals that skilled actors try to reach. In contrast to the model of objectivity that sees researchers *following* rules, we might see researchers variously *achieving* rules. When they have the right expertise, people can accomplish formal behavior, which allows for the coordination of actions. If this is right, then we can accept that formal products are essential to the objectivity and applicability of science, and also accept that expertise is necessary to achieve formal behavior.

Science and technology are extremely efficient at generating social order. We saw this in connection with the creation of order at the level of data (Chapter 10). Every time researchers come to agreement, whether it is about an observation, an interpretation, a phrasing of conclusions, or a theory, they are producing order. That order typically comes to be seen as the result of the order of nature or necessity. It only takes a little bit of study to see that nature and necessity do not routinely force themselves on scientists and engineers. Therefore the work that is done to make orders manifest is a central part of the study of science and technology (e.g. Turnbull 1995).

None of this is to say that formal rules and tools have mere fictional status. In addition to being goals they are also results. As results, people can build on these formal tools, in a way that they cannot build on more messy and context-laden accounts. It is these new powers that allow science and technology's successes to be more than isolated events or miracles.

Conclusions

Standardization and formal objectivity are things that people develop in the right social circumstances. Probably more importantly than in many other domains, scientists' and engineers' efforts to create objective procedures mean that formal tools are often entrusted with considerable responsibilities.

The ideal of formal objectivity runs counter to the demands of the messy world and the constant possibility of novel interpretations. So one question for this chapter was how to reconcile views of science that emphasize the strength of objective procedures in science and technology with the humanist models that emphasize expertise.

ANT and other models that emphasize objectivity claim that science's pattern of successes can only be understood as built on objective or mechanical procedures. When we look closely at supposedly objective procedures, though, we find that they are permeated with human skill at dealing with local contingencies. We should not see those procedures as produced by internal mechanical rules, then. Rather, objectivity is produced by human abilities to create order. We will thus always see expertise at work if we want to. But we can also abstract away from that expert work, trusting people to fit their actions to those objective procedures and to make formal tools capable of performing valuable work.

Science and technology produces social order. As with any group of people, every time scientists and engineers come to agreement about an observation, an interpretation, a phrasing, or a theory, they are producing order. Though nature does not routinely force its regularity upon researchers, science and technology's orders typically come to be seen as reflecting and caused by the order of nature. And that is why the work done by scientists and others to make the order of nature manifest is so interesting.

13

Rhetoric and Discourse

Rhetoric in Technical Domains?

Ancient Greece saw a number of rhetorical traditions and divisions arise, that have their intellectual descendants today (Conley 1990). The sophist Gorgias held that rhetoric was asymmetric, involving a skilled speaker trying to persuade or affect a passive audience. For Protagoras, rhetoric was the study of debate, allowing groups of people to examine various sides of a question. For Isocrates, rhetoric was about eloquence, and his educational program set out to create participants in civilized discourse. Aristotle's novel contribution was to see rhetoric as a tool of discovery, for generating important points. And opposed to all of these, Plato valued dialectic, in which the goal of argument is to establish truth with certainty, and not merely to persuade, explore, or be eloquent; Plato and many later descendants rejected "rhetoric" as unsavory (Conley 1990). Unfortunately, as various points so far in this book have shown, few arguments can establish truth with certainty. Moreover, Platonic arguments can be seen to depend upon an anti-democratic view that imagines most individuals as without knowledge or skill (Latour 1999). The arguments made in scientific and other technical contexts can sometimes be very persuasive, but to study how they are persuasive is the province of rhetoric.

Though there is no single scientific or technical style, we think of scientific or technical writing as typically dense, flat, and unemotional, precisely the opposite of rhetorical. Nonetheless, there is an art to writing in a factual and apparently artless style: "The author cannot choose whether to use rhetorical heightening. His only choice is of the kind of rhetoric he will use" (Booth 1961: 116). STS adopts this expansive claim for rhetoric, for a straightforward reason: every piece of scientific writing or speech involves choices, and the different choices have different effects. The scientific journal article is an attempt to convince its audience of some fact or facts. That means that every choice its authors make in writing an article is a rhetorical choice. The arguments, how they are constructed, the language of their construction, key terms, references, tables, and diagrams, are all selected for their effects. What is more, the scientific writing process often involves interplay among multiple authors, reviewers, and editors, making at least some choices and their rationales open to investigation. These points are even recognized by authors who want to insist on clearly separating rhetorical from epistemic issues (e.g. Kitcher 1991).

The fact that technical and scientific writing is not a transparent medium is apparent from the fact that dozens of manuals and textbooks on technical and scientific writing are published every year in English alone, the dominant scientific language. And scientific writing has not remained static. The history of the experimental report shows that it grew out of particular contexts, and was shaped by particular needs. When Isaac Newton wanted to communicate his theory of light, in his "New Theory of Light and Colours" of 1672, he recast a collection of observations to form a condensed narrative of an experiment (Bazerman 1988). The single experiment was in some sense a fiction, a result of Newton's writing, but the narrative form was compelling in a way that a collection of observations and pieces of theoretical reasoning would not be. Newton was participating in already established tropes of British science, and was refining the genre of the experimental report. Genre is by its nature an unstable category, because each instance of it affects the genre. The experimental report has changed and diversified considerably over the past 300 years, becoming adapted to new journals, new disciplines, and new scientific tools. Nonetheless, as it was for Newton, today's report is a narrative that seeks to convince the reader of a particular claim.

The Strength of Arguments

The point of most scientific discourse is ostensibly to establish facts. Though there are reasons to believe that aspects of scientific discourse are not primarily about facts, or at least not about the facts they overtly claim to be about, the establishment of facts is clearly central. Scientific articles are typically arguments for particular claims, experiments are usually presented as evidence for some more theoretical claim, and debates are about what is right.

What are the markers of a fact? Key markers include the lack of history or modality (Latour and Woolgar 1986). Facts are usually presented without trace of their origins and without any subordination to doubt, belief, surprise, or even acceptance. The best-established scientific facts are not even presented as such, but are instead simply taken for granted by writers and researchers in their attempts to establish other facts. The art of positive scientific rhetoric, then, is the art of moving statements from heavily modalized to less-modalized positions. The art is to shift statements like "It has never been successfully demonstrated that melatonin does not inhibit LH" to "Since melatonin inhibits LH . . ."

For Latour and Woolgar, key to science's rhetorical art is operations on texts. In their study of a duel over the structure of thyrotropin releasing factor (or hormone) – with a Nobel Prize at stake – they show how articles from two competing laboratories are filled with references to each other, but how those references involve skilful modalization and de-modalization (Latour and Woolgar 1986). Small markers of disbelief, questions, and assertions allow the two camps to create very different pictures of the state of the "literature" and the established facts.

Actor-network theory draws some of its concepts from the study of discourse, in that actors are identified by their being invoked in the course of arguments. A result of this is that the strength of an argument depends largely on the resources or *allies* that it brings to bear on the issue (Latour 1987). Citations are one source of allies: to cite another publication without modality, with a reference like ". . . (Locke 1992)," means that on that point David Locke will back the text up, that in his 1992 book he provides evidence to support the text's claim. Typical scientific arguments involve stacking allies in such a way that the reader feels isolated. The image here is of an antagonistic relationship between writers and readers. The scientific article is a battle plan, or "one player's strokes in [a] tennis final" (Latour 1987: 46). Facts and references are arranged to respond in advance to an imaginary audience's determined challenges. Arguments, then, are typically justifications of claims (Toulmin 1958). Good articles lead readers to a point where they are forced to accept the writers' conclusions, despite their efforts to disagree. In a provocative turn, Latour says that when an article is sufficiently rhetorical, employing sufficient force, it receives the highest compliment: it is *logical* (Latour 1987: 58).

For readers of scientific articles to challenge a claim requires breaking up alliances, achieved, for example, by challenging evidence or by severing the connection between the cited evidence and the claim. This process typically takes them further and further away from the matter at hand, finding either weak points in the foundation or finding contrary, facts too well established to be overturned. While foundationalism is an inappropriate philosophical picture of science (Box 2.2), it might be an appropriate picture of the rhetorical situation of scientific arguments.

For ANT, the allies brought to bear on a position are not just cited facts, but include material objects: the objects manipulated in laboratories, found in the field, surveyed, and so on. These objects are represented in scientific writing, and in some sense stand behind it. Just as readers attempt to sever connections between cited and citing articles, they attempt to sever the connections between the objects represented and the representations. If this cannot be done conceptually, through clever questioning, readers may find themselves led back to laboratories where they can try to establish their own representations of objects. In this way, ANT's rhetoric of science makes space for the material world to play a role.

The Scope of Claims

Equally important to the strength of arguments is their scope, the audiences and situations that they can be expected to address. Greg Myers's *Writing Biology* (1990) is a study of some related grant proposals, scientific articles, and popular articles (Myers (1995) has a similar analysis of patent applications). The histories of these proposals and articles, seen in multiple drafts of each, and reviewers' and editors' comments, show a series of rhetorical responses to different contexts. In particular, considerations of the identities and interests of readers shaped the writing of the different texts, producing works that were suited to promoting the authors' and reviewers' interests while catering to those of the readers.

Arguments have higher levels of *externality* the wider the scope of their claims: an article that simply describes data typically has a very low level of externality, but one that uses that data to make universal claims about wide-ranging phenomena has a high level (Pinch 1985). Prestigious interdisciplinary journals such as *Science* and *Nature* accept only articles that make claims with high levels of externality, but reviewers are also attuned to issues of overgeneralization. In addition, reviewers want to see articles that make novel claims, but ones that address existing concerns. And they want to see journal pages used efficiently, so only important articles can be long ones. Authors must simultaneously make novel claims of wide scope and support those claims solidly, in tightly argued articles that address existing concerns of readers.

The review process centrally involved adjustments of the level and status of each of the claims made by the articles. In both of Myers's (1990) cases, articles were originally rejected by the prestigious journals to which they were submitted, were resubmitted, rejected again, and eventually found their way to more specialized journals. The initial reasons for rejection centered on inappropriateness, because of the scope of the claims. The articles went through multiple drafts, ending in multiple articles addressing different audiences. The result of this process was the publication of articles that made different claims than did the original submissions, and were substantially different in tone.

Rhetoric in Context

The rhetorical analyses mentioned so far are studies of scientific persuasion in general, even while they are attached to specific texts. But persuasion needs to be attuned to its actual audience, and for this reason rhetorical analysis often needs to pay attention to how the goals of individual writers interact with norms of scientific writing and speaking.

In the broadest terms, there are two *repertoires* that scientists use, in different circumstances (Gilbert and Mulkay 1984). When discussing results or claims with which they agree, or when they are writing in formal contexts, scientists use an *empiricist* repertoire that emphasizes lines of empirical evidence and logical relations among facts: the empiricist repertoire justifies positions. When discussing results or claims with which they disagree, scientists use a *contingent* repertoire that emphasizes idiosyncratic causes of the results, and social or psychological pressures on the people holding those beliefs: the contingent repertoire explains, rather than justifies, positions. Since scientists move back and forth, there are reasons to challenge the priority of the empiricist over the contingent repertoire; both should be subject to discourse analysis (Gilbert and Mulkay 1984).

Aristotle's rhetorical *topoi* or *topics* are the resources available in particular contexts. For example, Mertonian norms are topics available to science as a whole: to challenge the legitimacy of a scientist's work one might employ disinterestedness, displaying how extra-scientific interests line up with scientific judgments (Prelli 1989). There are also rhetorical topics specific to disciplines or methods; for example, there are recurring difficulties in attempts to apply the mathematical theory of games, so a writer applying that theory has established topics to address, whether in political science or evolutionary biology (Sismondo 1997).

Rhetoricians might look at specific texts as more than examples (though also examples), seeing them as worthy of study in their own light. At least three prominent rhetoricians of science have scrutinized James Watson and Francis Crick's one-page article announcing their solution to the structure of DNA (Bazerman 1988; Gross 1990a; Prelli 1989). Stephen J. Gould and

Richard Lewontin's anti-adaptationist article, "The Spandrels of San Marco," is the subject of a book, representing about a dozen different rhetorical understandings of it (Selzer 1993).

When researchers are dealing with new or unfamiliar questions or phenomena, as opposed to answering a well-established question, they need to convince their readers that they are dealing with something real and worth paying attention to. The rhetorical task in these situations is to give presence to a question or phenomenon, to make it real and worthwhile in the minds of readers (Perelman and Olbrechts-Tyteca 1969). For example, taxonomists have to engage in considerable work to erect a new species (Gross 1990b). Not only do the taxonomists make arguments that their putative species is unknown, but they make their species salient in their readers' minds. Through pictures, charts, and a depth of observations, they create a unified phenomenon, available to readers.

Not all rhetoric is devised to establish facts. For example, Mario Biagioli's study *Galileo, Courtier* (1993) looks at Galileo's work, and in particular at a number of his public disputes, in terms of Galileo's professional career: he successfully rose from being a poorly paid, low-status mathematician to being a well-paid, high-status philosopher (or scientist). In seventeenth-century Italy, a key ingredient for a successful intellectual career was patronage, so Galileo's scientific work was often rhetorically organized to improve his standing with actual and potential patrons. Positions were articulated to increase not only Galileo's standing but that of his patrons. Even the very fact of the disputes is linked to patronage, in that witty and engaging disputes were a form of court entertainment, and contributed to the status of the court (Biagioli 1993; Findlen 1993; Tribby 1994).

Reflexivity

The study of rhetorics and discourses has sometimes pushed STS in a reflexive direction. Recognizing the ways in which a piece of scientific or technical writing or speech is rhetorically constructed invites attention to the rhetorical construction of one's own writing (e.g. Ashmore 1989; Mulkay 1989). All of the rhetorical issues identified so far are also issues for work in STS. Data has to be given definition, propositions have to be modalized and de-modalized, rhetorical allies have to be arranged, the scopes of claims need to be adjusted, appropriate topics have to be addressed, and problems have to be given presence. Facts in STS are no less rhetorically constructed than are facts in the sciences or engineering. Reflexive approaches thus explore the construction of facts *per se*.

One of the ways in which rhetorics can be displayed is via *unconventional* or *new literary forms*. By writing in the form of a dialogue or play, competing viewpoints can be presented as coherent packages, and disputes can be left unresolved. Having a second, third, or more voices in a text provides a convenient tool for registering objections to the main line of argument, for indicating choices made by the author, or for displaying ways in which points made can be misleading (Mulkay 1985). Unconventional forms draw attention to the shapes that authors normally give their texts, which hide the temporality of research and writing, the serendipity of topics and arguments, and the reasons and causes for texts appearing as they do. Unconventional forms, then, draw attention to the ways in which normal genres are conventional forms geared to certain purposes.

Reflexive approaches in STS have been the subject of some criticism. While they draw attention to conventional forms, and are often highly amusing to read, unconventional forms rarely change the relationship of the author and the reader significantly. Even while they appear to challenge arguments or claims, the challenges they present are in the control of the authors. Even while they appear to display temporality, serendipity, or background causes, they display these in the authors' terms. Finally, there is a tension between the attempt to establish a fact and the attempt to show the rhetorical construction of that fact, because the latter appears to delegitimate the former. Even though reflexive analyses typically show that STS is in exactly the same position as the fields it studies, they appear to show weaknesses rather than strengths. Reflexive approaches, then, are very useful for learning about general processes of fact-construction, but do not by themselves solve any problems, and may create their own rhetorical problems (see Collins and Yearley 1992; Woolgar 1992; Pinch 1993b; Woolgar 1993 for some interesting exchanges).

Metaphors and Politics

According to the positivist image, ideal scientific theories are axiomatic, mathematical structures that summarize and unify phenomena. In this picture, there is usually no important place for metaphors, which are viewed as flourishes or, sometimes, as aids to discovery, but they are never essential to the cognitive content of theories. However, almost every scientific framework depends upon one or a few key metaphors (Hesse 1966; Haraway 1976), and work in STS shows that scientific models and descriptions are at least replete with explicit metaphors and analogies. A number of works in the history of science and technology discuss particular metaphors, especially ones that have ideological value. We saw some of these in connection with feminist studies (Chapter 7); there are many other examples. For example, different narratives in primatology and in popularizations of primatology reflect important currents of thinking about men and women in Western societies, and also about relations to non-Western societies (Haraway 1989). Slightly differently, neo-classical economics appropriated the formalisms of then-contemporary physics of energy, resulting in a metaphorical connection between energy and value; nineteenth-century physics, then, shaped much twentieth-century economic research, in ways and directions of which the researchers have been largely unaware (Mirowski 1989). Descriptions of technologies are replete with metaphors that do political work. To take one example, in political contexts the description of the Internet as a highway positions it as infrastructure that serves the public good, that requires teams of experts, and importantly requires government investment (Wyatt 2004). This metaphor has helped to shape the present and future of the Internet.

For some analysts of science (e.g. Gergen 1986; Jones 1982) the ubiquity of metaphor even raises questions about science's claims to represent reality accurately. If metaphors are so prevalent, how can we say that scientific theories describe reality rather than construct frameworks? In particular, how can inquiry influenced by metaphors current in the wider culture, as so much of science is, be taken as representing?

Why are there so many metaphors? Metaphors in science are crucial as heuristic and conceptual tools (e.g. Hoffman 1985; Nersession 1988), and often serve important descriptive and referential functions (e.g. Ackermann 1985; Cummiskey 1992). The ubiquity of metaphor and analogy in the sciences can be taken as evidence that literal language lacks the resources for easy application to new realms (Hoffmann and Leibowitz 1991). Metaphors can define research programs rich with questions, insights, and agendas for research (Boyd 1979). They can become so rich that they become invisible – Lily Kay (1995) argues that the dominant metaphor of genes as information was in fact a contingent development, though now it is almost impossible to think of genetics except in informational terms.

Scientific metaphors can even police national boundaries. Agricultural research in the United States has on a number of occasions in the twentieth century established boundaries between "native" species and dangerous "immigrants." Washington's famous cherry trees were a gift from the Japanese government in 1912, but a precarious one. In a context that included anger and fear over (human) Japanese immigration, an earlier (1910) gift of two thousand cherry trees had been burned on the advice of the Bureau of Entomology, because they were variously infested (Pauly 1996). The later gift came along with strong arguments from Tokyo as to the health and safety of the trees. Some similar patterns can be seen in agriculturists' reactions to the kudzu vine (Lowney 1998).

Theories and models are abstractions, approximating away from the truth (Chapter 14). Too-tight correspondence to the world is something to escape from. At the theoretical level, scientists aim to elucidate the structures of material things. But abstractions have to take place within a framework, in the form of a lens through which to choose elements to abstract. Metaphors can provide such a lens, allowing ideology and truth to coexist.

Conclusions

Epistemic issues are simultaneously issues about persuasion. Even if our focus is narrowly on the production of scientific and technical knowledge, then, the study of rhetoric is crucial to understanding what people believe and how they believe it. Once that is granted, there are many different approaches to understanding persuasion, focusing on rhetorical force or contextually delimited topics.

There is also more to science and technology than the direct production of knowledge. Scientific and technical writing is tied to a variety of contexts, including the professional and personal contexts of scientists and engineers, questions of the legitimacy of approaches and methods, and broad ideological contexts. Often these contexts can be made visible only by the study of rhetoric, showing the metaphorical connections between discourses, goals, and ideologies.

14

The Unnaturalness of Science and Technology

The Status of Experiments

Experiments came into their own as sources of knowledge in the seventeenth century. Since then, they have tended to replace studies in the field where possible. This is not an obvious development. Considering the unnaturalness of experimental conditions, and their inaccessibility to other interested researchers, experimentation looks like a strange and fragile way of gaining knowledge about the natural world. However, science has largely adopted an "interventionist" approach to research (Hacking 1983). Thus experiments have been the focus of more attention in STS than have observations in the field (for a few exceptions see Clark and Westrum 1987; Bowker 1994; Latour 1999).

Most understandings of science view experiments as ways of deciding among theories. In the falsificationist view (Chapter 1), for example, theories are imaginative creations that stand or fall depending upon observations, typically laboratory observations. Theories, then, appear to have an independent scientific value that experiments do not have. In most scientific hierarchies, theories sit above experiments, and theorists sit above experimenters. Attention to laboratories, however, has started to erode the theory/experiment hierarchy in STS's view of science, if not in science itself. There are a number of reasons for this.

"Experiments have lives of their own," says Ian Hacking. Many experiments are entirely motivated and shaped by theoretical concerns. But at least some experiments are unmotivated by anything worthy of the name "theory." Hacking quotes from Humphry Davy's 1812 textbook, in which Davy provides an example of what should count as an experiment: "Let a wine glass filled with water be inverted over the Conferva [an algae], the air will collect in the upper part of the glass, and when the glass is filled with air, it may be closed by the hand, placed in its usual position, and an inflamed taper introduced into it; the taper will burn with more brilliancy than in the atmosphere" (quoted in Hacking 1983: 153). While Davy's actions in his example are shaped by ideas, there is no important sense in which they are the products of a discrete theory or set of theories. And Davy certainly does not set out to test any particular theory.

Once we accept that there can be experiments independent of theory, must we accept that experimentation is in general unrelated to theory? Surely, even experiments like Davy's are justified in terms of their potential contribution to theory? This claim is certainly right, but we should note that it amounts to little more than a reiteration of the theory/experiment hierarchy. We can read the slogan that experiments have lives of their own as saying that the hierarchy leaves space for a relatively autonomous activity of experimentation.

One of the innovations of STS has been to look at science as work. Particularly, those members of the field most influenced by the symbolic-interactionist approach to sociology have emphasized how science consists of interacting social worlds that are loci of meaning (e.g. Star 1992; Fujimura 1988; Casper and Clarke 1998). Participants are invested in their worlds, and attempt to ensure those worlds' continuance and autonomy. They strive to continue their own work and maintain identities. Looked at from the perspective of work, experimentation and theorizing are not hierarchically related, but are different (broad) modes of scientific work.

Knorr Cetina (1981) says that "theories adopt a peculiarly 'atheoretical' character in the laboratory": laboratory science aims at success, rather than truth. That is, the important goals of laboratory researchers are working experiments, reliable procedures, or desired products, rather than the acceptance or rejection of a theory. Knowledge is created in the laboratory, but "know-how" is primary. "Knowledge that" is largely a reframing of laboratory successes for particular audiences and purposes. In the laboratory theories take a back seat to more mundane interpretations, attempts to understand particular events or phenomena.

Finally, does theory contribute more to technology than experiment? It is not at all clear whether one is more important than the other. Theories have to be supplemented to be applicable. Students of technology have pointed out that scientific knowledge is often inadequate for the purposes of engineers, who may have to create their own knowledge (e.g. Vincenti 1990). The standard picture of technology as wholly dependent on science for its knowledge is flawed. Once we recognize that flaw, it is unclear whether theoretical knowledge or experimental know-how provides more help to engineers. Again, experimentation is not clearly or simply subordinate to theorizing.

Local Knowledge and Delocalization

Experiments are rarely isolated entities, but are rather connected to other experiments that employ similar patterns, tools, techniques, and subject matters. One way of grouping like experiments is by the notion of an *experimental system* (Rheinberger 1997). Teams often devote tremendous energy to developing a combination of tools and techniques with which to run any number of varying studies. For example, the "Fly-group" of geneticists led by T. H. Morgan in the early decades of the twentieth century developed *Drosophila* into a tool for studying genetics (Kohler 1994). To do so they bred the fly to better adapt it to laboratory conditions, and they developed techniques for further breeding and observing it. On the basis of that experimental system a minor industry in *Drosophila* genetics emerged. Similar stories can be told about the Wistar rat (Clause 1993), the laboratory mouse (Rader 1998), and other standardized laboratory organisms.

Laboratory work does not proceed in a purely systematic fashion (Chapter 10). Plans run into local resistance, which is dealt with locally. Materials do not behave as they are supposed to, equipment does not function smoothly, and it is difficult to configure everything so that it resembles the plan. The laboratory contains local idiosyncrasies that become central to day-to-day research. Researchers face a broad range of idiosyncratic issues, from employment regulations – which might "prohibit testing after 4:30 p.m. or at weekends" – to problems with materials – "the big variability is getting the raw material. We have never been able to get the same raw material again . . . You have to scratch yourself in the same place every time you play, and everything has to be the same, or else the accounts are meaningless" (quoted in Knorr Cetina 1981).

Given these issues, the problem of replication is a problem of delocalization. An experimental result occurs in a specific time and place. Scientific knowledge tends to be valued, though, the more it transcends time and place. But to de-localize or generalize a fact is a difficult task. Local features have to be divided between those crucial to success and those that are idiosyncratic. In addition, one has to identify what it is that is to be generalized, if anything. Knowing that dioxin was strongly linked to cancer in 20 laboratory rats, can one conclude that dioxin is carcinogenic in all mammals, in all rodents, in all laboratory rats, in all well-fed, under-stimulated animals . . . ? What class of things did the 20 laboratory rats represent?

The Unnaturalness of Experimental Knowledge

Scientists construct systems in order to escape the messiness of nature, to study controlled, cleaned, and purified phenomena about which models can be more easily made. Experimentalists can know much more about artificial phenomena than messy nature, so they rarely study nature as it is. Nature is systematically excluded from the lab (Knorr Cetina 1981; Hacking 1983; Latour and Woolgar 1986; Radder 1993 challenges the importance of this).

Experimental subjects and the contexts in which they are placed may be entirely artificial, constructed, or fabricated, making our knowledge about them not knowledge about a mind-independent world. The world of experiments, both material and thought experiments, can exist in the absence of naturalness – scientists can construct and investigate ever more simple or complex systems, according to their interests. Experimental knowledge is not knowledge directly about an independent reality, but about a reality apparently constructed by experimenters. In laboratories, where most experiments are performed, researchers do not need to accommodate natural objects and events as they are, where they are, or when they happen (Knorr Cetina 1999).

Put differently, experimental knowledge takes an enormous amount of work. As we have seen, it takes work to set up experimental systems that perform reliably. Those systems are scientists' creations, having been purified and organized so that they do not behave as chaotically as nature, and so that they do what the researchers want them to do. But experimental systems are rarely perfectly reliable. They do not necessarily respond well to every situation or set of inputs that experimenters wish to explore, so it also takes work to continually maintain the order of the laboratory. As we have seen, scientists attribute the order that they discover to nature, and the disorder to local idiosyncrasies or their own inadequate control of circumstances. But, as we have seen, there is a case for attributing science's ordering of nature not to nature but to scientific work itself (Chapter 10).

The artificiality of experiments was one of the concerns that many natural philosophers of the seventeenth century had about them. According to the scholastic ideal that dominated knowledge of the early seventeenth century, the natural phenomena that scientific knowledge studied were supposed to be well-established or readily observable regularities, like tides, weather, or biological cycles (Dear 1995a; Shapin and Schaffer 1985). Experiments could be demonstrative devices, but their artificiality prevented them from standing in for nature, and prevented them from establishing phenomena. Before they could become legitimate, experiments had to produce facts that would be accepted socially as opposed to individually; at that point they could properly display natural facts (Garber 1995). Before they could become revealing of nature, experiments had to become seen as potentially replicable by all competent experimenters, thereby fulfilling a norm of universality. Such concerns were addressed through the creation of an analogy between mathematical constructions and material ones: mathematical constructions enable learning about mathematical structures in the same way that material constructions enable learning about (natural) material structures (Dear 1995b). Particular places and spaces that served as laboratories contributed to the legitimacy of experiment – for example, the location of laboratories within the homes of English gentlemen helped establish trust in experimenters and the phenomena they displayed (Shapin 1988). And early experiments needed to be made into quasi-public events, even when they were done in private: Robert Boyle's detailed writing allowed for *virtual witnessing* of his experiments with air pumps (Shapin and Schaffer 1985). This helped to overcome the difficult burden of arguing that they were not idiosyncratic, private events, but publicly available reflections of natural regularities.

The Unnaturalness of Theoretical Knowledge

Experimentation's artificiality has parallels in the world of theory. Although we may think of theories and models as straightforward descriptions of the material world, identifying its states and properties, they are at best idealizations of or abstractions from the real world that they purport to represent. This point is at least as old as Plato's argument that mathematics must describe a world of forms that are prior to material reality; mathematicians in Ancient Greece developed styles of diagrams and technical language that allowed them to learn about necessary and abstract objects while working with constructed and concrete ones (Netz 1999). But this point has also been forcefully made more recently. Scientific theories and laws are *ceteris paribus* statements: they do not tell us how the natural world behaves in its natural states, but rather that a set of objects with such and such properties, and only those properties, would behave like so (Cartwright 1983). By science's own standards, its best theories and laws rarely even apply exactly to the contrived circumstances of the laboratory.

Nancy Cartwright's (1983) close examination of physical laws reveals the artificiality of theories and laws (see also Box 14.1). She argues that physical laws apply only in appropriate circumstances. To take a simple example, Newton's law of gravitation is: $F = Gm_1m_2/r^2$. This states that the force

---

Box 14.1   Idealized islands, idealized species

Biologists R. H. MacArthur and E. O. Wilson's theory of island biogeography (IB) predicts the number of species (within some taxonomic group) that will be found on an island, on the assumption that immigration and local extinction will eventually balance each other out. The level at which they balance depends upon two variables: (i) the size of the island, because small islands have higher extinction rates, and (ii) how far it is from sources of colonization, because distant islands receive fewer immigrants. The theory was an important one in ecology, and the debate around it shows conflicting views of abstraction and idealization (Sismondo 2000).

IB was often viewed as obviously right at a high level of abstraction. To most theoreticians the truth of IB was not in question, even though its predictions were not very good, and even though it made assumptions that are not strictly true. But for many empirically minded ecologists the situation was the reverse. The data did not support IB, and the theory made key assumptions that were too false to accept: it assumed that all the species in question were essentially identical, and that all the islands in question were essentially identical except for size and location.

In the end, the debate was one over the relative status of theoretical and particular explanations. To the extent that ecologists were able to create particular explanations, and interested in creating them, they found IB too abstract to count as possibly true. To take a popular example, according to the theory the extinction of the dodo was made more likely by its being located only on small islands isolated in the middle of the Indian Ocean. Yet while the theory explains this extinction, it does so in an unsatisfying way. The explanation discards too much information, information about, for example, the birds' naiveté, the explosion of human travel since the sixteenth century, and the voraciousness of modern hunters. In historical terms, we know what happened to the dodo, and the theoretical terms offered by IB seem pale in comparison. To empirically minded ecologists, nature is a collection of particular events. In the process of investigating that nature they get their boots dirty or otherwise immerse themselves in data, and IB fails because it fails to attend closely enough to those particulars.

---

between two objects is directly proportional to the product of their masses. But this is only right if the two objects have no charge, for otherwise there will be an electrical force between them, following Coulomb's law, which is similar in form to Newton's. We can correct these two laws by combining them, but at the expense of their usefulness in the spheres in which they are traditionally used. Or we can divide the single force acting between two bodies into two components, a gravitational force and an electrical one. Thus, "we can preserve the truth of Coulomb's law and the law of gravitation by making them about something other than the facts" (Cartwright 1983).

For explanatory purposes, scientists want widely applicable laws, even at the expense of their being precisely applicable in any real case. They move the discussion away from the straightforward facts of nature, making their theories about underlying structures of nature. Like experimental knowledge, theory is about cleansed and purified phenomena, abstractions away from the truth.

Similarly, models adopt simplifying assumptions that reveal key features of their objects. Good models should be unobvious but intuitively right, at least once one accepts their abstractions. They are created with solutions in mind, so that the modelers can make definite statements about the situations they describe (Breslau and Yonay 1999; Boumans 1999). At times, these simplified and artificial models can help to make themselves true. Game theoretic models of nuclear war shaped actual strategic thinking, when sides assumed that others were engaging in game theory (Ghamari-Tabrizi 2005). Models of finance have shaped actual markets, when, for example, the Black–Scholes Equation and the related Capital Assets Pricing Model allowed for the creation of a huge derivatives market; the title of Donald MacKenzie's book on finance theory, *An Engine, Not a Camera* (2006), provides its central thesis.

Therefore, theories, models, and by extension even many isolated facts, do not mirror nature in its raw form, but instead may describe particular facets or hidden structures, or may be attempts to depict particular artificial phenomena.

A Link to Technology?

As we saw in Chapter 9, it is not obvious that there should be a close connection between science and technology, and in fact the two stand somewhat further apart than many people assume. At the same time, science contributes substantially to technology, and the two seem increasingly entangled (Latour 1987; Gibbons et al. 1994; Stokes 1997). The unnaturalness of experiment suggests a reason why science and technology are as linked as they are.

If experimental science provides knowledge about artificial realities, it, and sometimes theorizing and modeling, provides knowledge about the

Box 14.2   Genetically modified organisms

Nature, of course, has immense symbolic value. Technological controversies can easily develop around the acceptable limits of the unnatural. Debates over antidepressant drugs, new reproductive technologies, and genetically modified foods often revolve around the need for natural emotions, childbirth, and food.

Genetically modified (GM) foods are at the center of the debate over genetic engineering more generally. The number and variety of highly public issues that critics have raised is astonishing, suggesting deep-seated aversions to the idea of GM foods. Could GM foods provoke unexpected allergic reactions? Could foreign gene sequences get transferred to microorganisms in people's intestinal tracts? Do GM foods contain inedible and therefore harmful proteins? Will dangerous genes spread when GM crops hybridize? Will GM foods threaten the food chain when insects eat them? Will non-targeted insects be threatened by insecticides produced by GM plants? Will non-targeted plants be threatened by herbicides used on herbicide-resistant crops? Will herbicide use increase as a result of herbicide-resistant crops? Will GM crops simply increase the dependence of farmers on agribusiness? Will they lead to more consolidation of farms? – Proponents have, unsurprisingly, made an equally large number of positive claims for GM foods.

Anti-GM alliances are relatively unified, even when they involve somewhat contradictory positions, again suggesting deep-seated concerns. Pro-GM writers invoke consumer rights when GM foods are at issue, and anti-GM writers invoke consumer rights when the *labeling* of GM foods is at issue; similarly, pro-GM writers emphasize the flexibility and uncertainty of knowledge behind labels, while anti-GM writers emphasize the flexibility and uncertainty of knowledge behind the technologies themselves (Klintman 2002).

Within scientific and technical communities, the loud public debates have their effects. To take one from many possible examples, when researcher Arpad Pusztai announced on television that GM potatoes had negative effects on rats' intestines, he was roundly criticized, and lost his job, for having circumvented peer review (see also Delborne 2008). An article on the study was eventually published by the medical journal *The Lancet*, but before it appeared the British Royal Society had already publicly pronounced it flawed, and the journal as biased. This provoked counter-charges from the editor of *The Lancet*: the Royal Society, he claimed, had staked its financial future on partnerships with industry, and as a result it and its members had been captured by industry interests. Because GM controversies are so public, scientific studies on the topic often provoke strong normative criticism, for being interested or otherwise violating the scientific ethos. Scientific realms, then, are not immune to concerns over the technological encroachment on nature.

structure of what can be done. Experimenters make systems in the laboratory, and investigate what those systems can do. Those systems may be very informative about the order of nature, but at root they are artificial systems. Engineers, too, create artificial systems, not usually with the goal of finding out about the natural world, but with the goal of doing some concrete work.

Given the gaps between science and nature, obvious questions arise about the possibility of materially bridging those gaps. How, if science does not describe nature, is science applied to nature? At the same time that it suggests a context for understanding the closeness of science and technology, the unnaturalness of both experimental and theoretical knowledge suggests a problem for understanding the success of applied science. Applied science has to succeed not because the knowledge on which it draws is right, but because of a mutual accommodation between the science and the settings to which it is applied (Latour 1988).

A different kind of link between technology and the unnaturalness of science might give us pause, when technologies depend upon complex laboratory work, and are used to provide representations of people. Laboratory conditions had to be created to make sex hormones visible: "sex hormones are not just found in nature" (Oudshoorn 1994). Instead, they are found in particular contexts, in which they can be made to act. They become seen as universal and natural because of "strategies of decontextualization" which hide the circumstances in which they can appear. Once naturalized, sex hormones can be used to help naturalize gender, and to affect it, through drugs. Similarly, positron emission tomography (PET) and magnetic resonance imaging (MRI) devices are not transparent devices for seeing into brains and other bodily tissues (Dumit 2003; Joyce 2008). Each presents the appearance of being a mere tool of visualization, even though the pathways by which they create images are much more complex. In their development, each depended on challenging work to turn it into a stable tool that presents that appearance. In their uses, and the readings of the images they create, they are not straightforward, but require skill and judgment. Yet they are taken, in medical and even in court contexts, to display persons as they really are.

The Order of Nature?

What is scientific knowledge about? It is about all types of things, from macro-evolutionary tendencies to genetic mechanisms, from black holes to super-strings, from the global climate to turbulence. We might be inclined to say that scientific knowledge is about natural objects and processes, and also a few particularly interesting artifacts. At the same time, most of the tools that produce scientific knowledge place it one or more removes from such objects and processes. The best scientific knowledge does not straightforwardly consist of truths about the natural world, but of other truths.

When psychologists try to understand the structure of learning they typically turn to experiments on rats in mazes, or on people in constrained circumstances, but they more rarely try to study the everyday learning that people do in complex ordinary environments. Knowledge stemming from experiments most directly describes situations that are distinctly non-natural, standing apart from nature in their purity and artificiality. Experimenters make sure that their inputs are refined enough to be revealing, and use those inputs to create relationships not found outside the laboratory. They learn about something that they see as more basic than the immediate world. Similarly, science's best theories don't describe the natural things we observe around us, or even the invisible ones that can be detected with instruments more subtle than our senses. Instead, those theories describe either idealizations or other kinds of fundamental structures. Few basic theories literally or accurately describe anything concrete, according to science's own standards. And computer simulations and mathematical models create new symbolic worlds that are supposed to run parallel to the more familiar ones, but are in many ways distinct: simulations are typically explorations of what would happen given specific assumptions and starting conditions, not what does happen. Even field science necessarily creates abstractions from the natural world, depending upon systems of classification, sampling, and ordering; in these ways it can be seen to project laboratories onto the field.

Most scientific knowledge both is and is not universal. It is universal in the sense that in its artificiality and abstraction it is not firmly rooted to particular locations. Theoretical knowledge is about idealized worlds; laboratory knowledge is created so that it can be decontextualized, moved from place to place with relative ease. Yet scientific knowledge is not universal because its immediate scope is limited to the artificial and abstract domains from which it comes – though there is always a possibility of its extension. Bruno Latour has an analogy that is useful here:

> When people say that knowledge is "universally true," we must understand that it is like railroads, which are found everywhere in the world but only to a limited extent. To shift to claiming that locomotives can move beyond their narrow and expensive rails is another matter. Yet magicians try to dazzle us with "universal laws" which they claim to be valid even in the gaps between the networks. (Latour 1988: 226)

Locomotives are very powerful, but they can only run on rails. Scientific knowledge is similarly powerful, but when it leaves its ideal and artificial environments it can quickly become weak.

Very roughly, we can understand the objects of scientific knowledge in one of two ways:

1. In constructivist terms – they can be seen as constructions of the researchers, as those researchers transform disorderly nature into orderly artifacts. Theoretical and laboratory tinkering shapes nature so that it can be understood and used. Order is imposed upon nature by science.
2. In realist terms – they can be seen as revealing a deeper order, which is absent in surface manifestations of nature. Scientists tinker to reveal structure, not to impose it. Science is an activity that discovers worlds that lie beneath, or are embedded in our ordinary one.

With some philosophical work, it is possible that these positions can be somewhat reconciled. On the one hand, constructions use available resources, and so depend on the affordances of those resources. Thus even constructions reveal something like a deeper order. On the other hand, given that there is no view from nowhere, reality is not a self-evident concept and may ultimately refer to the outcome of inquiry. What counts as a deeper order will depend upon what scientists and others can construct. This bare-bones sketch suggests that the difference between the above realism and constructivism may be a difference in emphasis.

For the most part, STS adopts constructivism over realism. However, in so doing the majority of scholars in STS have not adopted a magical or fanciful view. Rather, they believe that the mundane material actions of scientists in the laboratory produce new objects, not pre-existing ones.

15

The Public Understanding of Science

The Shape of Popular Science and Technology

We have seen that scientific and engineering research is textured at the local level, that it is shaped by professional cultures and interplays of interests, and that its claims and products result from thoroughly social processes. This is very unlike common images. It is very unlike, for example, the views of science and technology with which this book began. And it is very unlike the views one finds in popularizations. Why is this?

Although STS does not necessarily provide strong grounds for skepticism, the humanization of science and technology undermines some sources of scientific and technical authority. Given a humanized science we cannot say, for example, that scientific knowledge *simply* reflects nature, or even that it arrives from nature as the result of a perfectly rigid series of steps, so its authority cannot stem from nature alone. We cannot say that technologies *simply* unfold naturally and inevitably, and so the authority of engineers cannot stem from a too-easy narrative of progress. Claims about the constructedness of knowledge and technologies come up in science and engineering, but only in very specific contexts. For example, they come up when scientists and engineers use intellectual property law to defend their interests; then they need to argue for the constructedness of their knowledge in order to assert their creative contribution to it (McSherry 2001; Packer and Webster 1996). Claims about constructedness also come up when disputants in a controversy use a "contingent repertoire" to undermine their opponents' authority (Gilbert and Mulkay 1984). For this reason, the observations made in STS look threatening to scientists and engineers.

Journalism informed by STS, or similarly adopting a symmetrical approach, would not be popular with scientists. Science journalists are very closely allied with scientists (Nelkin 1995; Dornan 1990). All journalists depend upon their contacts for timely information, and science journalists are no exception. They

Box 15.1   The manufacture of ignorance

The discrepancy between the idealized version of science common in the media and more messy accounts can sometimes be used for particular purposes, establishing ignorance rather than knowledge (e.g. Stocking and Holstein 2008).

A number of interests are opposed to taking any measures in response to the threat of climate change. In the 1990s, one way in which that opposition was made effective was by questioning the phenomenon itself. Opponents of action drew on the fact that the scientific community did not speak with one voice on the issue (Edwards 1999). Given the standard image of science, to many people critics of global warming appeared more scientific than their colleagues, even though they stood outside of the consensus (e.g. Boykoff and Boykoff 2004). The aggressive campaign, especially in the United States, against the idea of climate change was helped along by veterans of earlier attempts to shape public opinion and sow doubt, in such areas as the Reagan Administration's Strategic Defense Initiative and the tobacco industry's attempt to confuse the connection between smoking and cancer (Oreskes and Conway 2008).

The tobacco industry was a pioneer in the creation of ignorance, as its slogan "Doubt is our product" suggests. As evidence became apparently incontrovertible that smoking caused lung and other cancers, the industry systematically argued that correlations could not show causation. Its legal strategy depended on combining the claim that it was common knowledge that smoking was dangerous and the claim that the scientific evidence was uncertain. For public relations purposes the industry also funded scientific research to challenge the connection, and to study other possible causes of cancer, from urbanization to sick building syndrome (Proctor 2008; Murphy 2006). And more recently, the industry has funded historical work to defend claims about earlier scientific ignorance.

Ignorance, then, can be a strategic resource, and therefore we might usefully develop social theories of ignorance (Smithson 2008), or study "agnotology" more generally (Proctor 2008).

thus participate in both informal and formal networks. Some important scientific journals, like *Nature, Science,* and the *New England Journal of Medicine,* send out advance copies of articles to select writers, on the condition that those writers not publish anything about those articles before the journal does (Kiernan 1997). These advance copies can allow science writers to have their stories, complete with interviews and quotes, prepared and waiting, and allow the journals timely publicity. In addition, to an extent that other journalists do not, science writers depend on their contacts for accurate facts and background information. Science is often esoteric and difficult to understand, and accuracy is a key value in writing about science, which amplifies the dependence on contacts. Science reporting in newspapers and magazines, and science writing in books more generally, may be shaped by these associations.

Perhaps most importantly, common styles of science journalism emphasize findings and their importance, not processes (e.g. Gregory and Miller 1998). Newspaper and other editors are interested in stories that convey excitement. Those stories tend to be about the definitive – at least at the top of the story – discovery of the biggest, smallest, or most fundamental of things. Doubts, questions, caveats, and qualifications are downplayed, sometimes to the dismay of scientists. Most readers are left with the impression of one or a very few researchers making substantial advances, and the rest of science immediately agreeing. Readers have come to expect this, and therefore one key part of popular science writing is usually an idealized description of the genius and logic behind a new discovery. The other key part is a description of the wonder of nature that has been revealed – wonders are just as crucial to good science writing as they are to tabloid pseudo-science. Popular science creates a "narrative of nature" (Myers 1990).

The Dominant Model and Its Problems

A number of scholars have recognized a "dominant model" (Hilgartner 1990), "canonical account" (Bucchi 1998), or "diffusionist model" (Lewenstein 1995) of science popularization: Science produces genuine knowledge, but that knowledge is too complicated to be widely understood. Therefore there is a role for mediators who translate genuine scientific knowledge into simplified accounts for general consumption. From the point of view of science, however, simplification always represents distortion. Popularization, then, is a necessary evil, not to be done by working scientists still engaged in productive research – the culture of science heavily discourages scientists from adopting the role of mediators, and shapes how they mediate when they do. Popularization pollutes the sphere of pure research.

Yet, popularization of science often feeds back into the research process (Hilgartner 1990; Bucchi 1998). In the cold fusion saga, because the claim to have created cold fusion first appeared in the media, most scientists who later became involved in the controversy found out about the claim on the same day as everybody else. The particular way of framing the claim about cold fusion, and something of its cachet, came from media reports (Bucchi 1998; Collins and Pinch 1993). Popularizations affect scientific research because scientists read them, and may even be a large portion of the "popular" audience. Even within specialized fields, scientists may cite articles more often if reports about them have been published in newspapers (Phillips 1991). Thus there is continuity between real and popular science, though scientists treat these as completely distinct. Popularizations may also affect the shape of scientific research, when they affect public and policy-makers' attitudes toward areas of research. For example, in the 1980s and 1990s, a number of planetary scientists promoted the idea that possible impacts of large asteroids posed a significant threat to Earth. Their work sparked and then drew on science fiction novels and films, which helped to create a narrative on which nuclear weapons in space would be heroic saviors of the planet (Mellor 2007). They contributed to a continued justification for nuclear weapons research and the militarization of space, as planetary defenses against asteroids.

The dominant model captures what scientists see as appropriate for normal science situations (Bucchi 1998). When disciplinary boundaries are well established and strong, then novel findings and ideas are easily dealt with, either by being incorporated into the discipline or rejected. Popularization follows a standard path, or is easily and clearly labeled as deviant. When disciplinary boundaries are weak, however, scientists may use popular media as an alternative form of communication; they may play out disagreements in the public eye, and even negotiate the science/non-science boundary there. In such cases, disciplinary resources are not enough to resolve conflicts, and so the outside media become more appropriate (Bucchi 1998). Although this account appears to capture one of the factors affecting the use of popular media in science, there are some examples in which researchers shape fields on the basis of powerful popularizations, such as Richard Dawkins's *The Selfish Gene* (1976).

Scientific rules about popularization are often applied self-servingly. Rules "are stressed by scientists who want to criticize or limit other scientists' behavior but are ignored by the same scientists with regard to their own behavior" (Gregory and Miller 1998). In addition, "scientists who do not popularize tend to see popularization as something that would damage their own career; however, they also think that other scientists use popularization to advance their careers." The dominant model is a resource used by scientists when it suits their purposes, and ignored when it does not.

Box 15.2   The Hwang affair

In 2004, Woo Suk Hwang, a prominent scientist at Seoul National University, published a paper in the journal *Science* announcing the first cloned human embryonic stem cell line. Another paper in *Science* followed a year later announcing 11 more cloned stem cell lines. Hwang had already gained national attention for earlier cloning work, and proved mediagenic in his modesty, his work ethic, his religiosity, and his nationalism, all highly valued in South Korea. Hwang became a national hero, and the South Korean government named him its "Supreme Scientist." The country had invested in biotechnology, identifying that as an area for new growth, important given increased competition in information technology and manufacturing (L. Kim 2008). Hwang came to symbolize the hopes for South Korean science.

In 2005 a whistleblower from Hwang's lab approached producers of the television show *PD Notebook*, a respected South Korean investigative journalism show, with a suggestion that the second paper might have been based on faked data. *PD Notebook* had already been quietly building a case that the human eggs for Hwang's research had been collected unethically, and used their earlier work as a starting point for the fraud investigation.

The controversy that ensued involved a clash of authority and power between Hwang and *PD Notebook*. Hwang challenged the journalists' competence and authority to investigate his work, in light of its validation by publication in *Science*. In response to charges of unethically procuring eggs from workers in his lab, Hwang charged his accusers with unethical journalism, because to obtain information they had lied to laboratory workers about how much they knew. But most importantly, Hwang mobilized supporters to start a mass boycott of products advertised on *PD Notebook*, and almost managed to force the show off the air before it had completed its investigation (J. Kim 2009). Only a university investigation, concluding fraud, saved the show and the television network.

Even after Hwang admitted misconduct, large numbers of supporters from among the general public continued to support him, showing strong emotional investment in his stem cell work. Rallies in his support continued, and substantial Internet-based groups continued to discuss the case, expressing their love for Hwang, their hopes for his research, and blaming various parties for the conspiracy against him (J. Kim 2009). Women continued to feel very positively about donating their eggs, and continued to do so, though in part because there is a donation culture in South Korea (Leem and Park 2008).

Around the world, excitement about the field, and to some extent its funding, had depended on promises of near-miraculous cures, when stem cell technology would allow for the re-growth of specific tissues. Hwang's initial papers had been widely reported on, had been seen as vindications of the field, and had contributed to expectations for the field. When the fraud was announced, researchers and journalists did careful boundary work (Chapter 3) to maintain hope for the field. They separated Hwang's particular research from promise of stem cell research, blaming the failure on distinctive styles of Korean science, and distinguishing different types of stem cell research (Kitzinger 2008).

The dominant model is a resource for more than just individual scientists, and can be seen as an ideological resource for science as a whole. The notion of popularization as distortion can be used to discredit non-scientists' use of science, reserving the use for scientists. In effect this enlarges the boundaries of the conceptual authority of scientists. This is despite the fact that science depends upon popularization for its authority. If there were no popularizers of any sort, then science would be a much more marginal intellectual activity than it is.

Perhaps most importantly, although scientists routinely complain about simplifications and distortions in popular science, they recognize other forums in which simplification and distortion is acceptable. At some level, most steps in the scientific process involve simplifications: descriptions of techniques are simplified in attempts to universalize them, the complexity of data is routinely simplified in attempts to model it, and so on (Star 1983). When outside researchers use results from a discipline or problem area, they routinely summarize or reshape those results to fit new contexts (Hilgartner 1990). Although particular cases of this reshaping are seen as distortions, in general it is accepted as legitimate. No sharp distinction can be drawn, then, between genuine knowledge and popularization: "Scientific knowledge is constructed through the collective transformation of statements, and popularization can be seen as an extension of this process" (Hilgartner 1990). Any statement of scientific knowledge is more or less well suited to its context, and might have to be changed to become suited to some other context, even another scientific context. Popularization is a way of moving knowledge into new domains. The popular context creates its own demands, and will tend to shape popular science accordingly.

The dominant model of popularization assumes that scientific knowledge is not tied to any context. Popularization pollutes pure scientific knowledge by simplifying or otherwise changing it to fit non-scientific contexts. That model ignores ways in which pure science can be continuous with its popularizations, ways in which "pure" science can even depend upon "popular" science to further research. The model also ignores ways in which science is historically located in disciplinary and other matrixes. Knowledge claims are contextualized and recontextualized within pure science, in ways that scientists understand and accept: a claim appropriate for cancer research may have to be reframed for a physiologist.

Interestingly, scientists can find themselves rejecting the dominant model, if they are marginalized and choose to adopt the stance of the dissident. Dissident scientists may challenge norms of science that emphasize its independence and self-sufficiency, instead seeing science as corrupted or permeated by politics, and seeing the public as a potential participant, corrective force, and source of accountability (Delborne 2008).

The Deficit Model

Going hand in hand with the dominant model of science popularization is the "deficit model" of the public understanding of science (Wynne 1992). On that model, scientific and technical literacy is a good in short supply outside the ranks of scientists and engineers. The public is thus characterized as deficient in knowledge. Shocking statistics about the number of people who believe that the Sun goes around the Earth and not vice versa, or that the Earth is less than 10,000 years old, easily motivate the deficit model.

Any deficiency must be seen as a problem. Given the centrality of science and technology to the modern world, scientific illiteracy is viewed as a moral problem, leaving people incapable of understanding the world around them and incapable of acting rationally in that world. For scientists, the deficiency also represents a political problem, because (presumably) the scientifically illiterate are less likely to support spending on science and more likely to support measures that constrain research. Therefore, many people feel that we need more "public understanding of science": the problem is one to be corrected by didactic education, a transfer of knowledge from science to broad publics.

While there is considerable sympathy within STS for the idea that publics should know more about science and technology, STS's perspective on expertise leaves it skeptical of the idea that the goal should be simply to teach people more science (e.g. Locke 2002). As a result, in STS the phrase *public understanding of science* often refers to *studies* of attempts to apply scientific knowledge or methods to problems in the public sphere. Here lay reactions to experts are as much of interest as experts' strategies for applying their knowledge.

Publics may have more nuanced relationships with scientific knowledge than the deficit model assumes. We have already seen that the dominant model of popularization mistakenly removes science and popularization from their contexts. The deficit model similarly fails to appreciate the contextual nature of knowing. Publics have knowledge that intersects with science, they may translate and appropriate scientific knowledge, and they appraise scientific knowledge and its bearers.

Because members of the public have pre-existing interests in certain problems and their solution, case studies often show a certain level of conflict between lay and scientific understandings. Steven Yearley (1999) summarizes the findings of these case studies in terms of three theses:

1. A large part of the public evaluation of scientific knowledge is via the evaluation of the institutions and scientists presenting that knowledge.
2. Members of the interested public often have expertise that bears on the problem, which may conflict with scientific expertise.
3. Scientific knowledge contains implicit normative assumptions, or assumptions about the social world, which members of the public can recognize and with which they can disagree.

Scientific knowledge is invariably at least partly tied to the local circumstances of its production. Science is done in highly artificial environments (Chapter 14); while these environments support versions of universality and objectivity (Chapter 12), they are nonetheless limited. When experts attempt to take science into the public domain, and are confronted by the interested public, those limits can often be seen, especially if the opposition is determined enough (Chapter 11).

In these public sphere cases, then, opposition to science is not the result of mere "misunderstandings," but of inadequate scientific work. If a proposed study of or solution to a public sphere problem is not put forward by trustworthy agencies or representatives, fails to take account of lay expertise, or makes inadequate sociological assumptions, then it may encounter opposition grounded in legitimate concerns.

Brian Wynne's study of Cumbrian sheep farmers (Box 15.3) shows us how these lessons play out in practice (the history of AIDS treatment activism of Box 16.1 can be used to make similar lessons). In terms of Yearley's theses:

Box 15.3 Wynne on sheep farming

Following the nuclear accident at Chernobyl in late April 1986, a radio-active cloud passed over northern Europe. Localized rainstorms dumped caesium isotopes on various parts of Britain, particularly at higher elevations. The British government's initial reaction was to dismiss the contamination as negligible, but in mid-June a three-week ban was placed on the move-ment and slaughter of the affected sheep in one region, Cumbria. At the end of the ban, the restrictions were made indefinite (Wynne 1996).

The scientific advice to the farmers was to do nothing, because the sheep would decontaminate themselves quickly as the caesium sank into the soil and fresh uncontaminated grass grew. It turned out that this advice was based on incorrect assumptions about the soil, only discovered two years later. Meanwhile the farmers' attention turned to the Sellafield Nuclear Plant, which sat approximately in the center of the affected areas. Although there had been uneasiness about Sellafield for years, that uneasiness was countered by community ties to the plant, the largest employer in the area. Now the farmers had more concrete reason to believe that the plant was doing harm. The main concern about Sellafield involved a 1957 fire, which had emitted a lot of radioactivity. As became clear in the years following Chernobyl, the dangers caused by the Sellafield fire had been covered up, and the fire itself had provided an opportunity to cover up routine emis-sions of spent fuel. Sellafield had also been at the center of controversies and criticism in years leading up to Chernobyl, because of elevated rates of leukemia, accusations of illegal discharges, apparently misleading informa-tion given to inquiries, and poor safety management.

The sheep farmers developed a thorough skepticism of government posi-tions and those of their scientific representatives. They did not believe the claims of scientists from the Ministry of Agriculture, Fisheries and Food that the caesium "fingerprint" in the Cumbrian hills matched that of Chernobyl rather than Sellafield; later studies suggested that the farmers were right, that 50 percent of the radioactivity did not come from Chernobyl. When the Ministry advised farmers to keep their lambs in the cleaner valleys longer, that advice was ignored, because the farmers knew that the valleys would become depleted. An experiment in decontamination involved spreading bentonite at different concentrations in different fields, and fencing in the sheep in those fields; farmers objected that sheep do not thrive when fenced in, but they were ignored, and the experiments were eventually abandoned precisely because the fenced-in sheep were doing poorly.

1. The farmers had a history of muted distrust of government positions and of government scientists on the issue, because they had consistently downplayed dangers, because they appeared to have been covering up problems, and because they made errors about matters on which they claimed expertise.
2. On a number of occasions scientists ignored the farmers' own expertise about the habits of sheep and the productivity of the hillsides.
3. Scientists made assumptions about the culture and economics of sheep-farming, assumptions that disagreed with the farmers' knowledge of themselves.

The farmers can be seen to have a reasonable response to the scientists, based in a culture that is quite divergent from the scientific one. Opposition to science was not the result of a misunderstanding, but was the result of inadequate trust and connections between scientific and lay cultures with very different knowledge traditions.

STS's challenge of our common view of science and technology has some interesting consequences when, as it is here, it is brought to bear on the boundaries of scientific knowledge. The dominant model of expertise assumes that science trumps all other knowledge traditions, ignoring claims to knowledge that come out of non-science traditions. For this reason, when scientists find opposition to their claims, they tend to see that opposition as misinformed or even irrational. Moreover, controversies between experts and non-experts may not be resolved in the way that controversies among experts are: mechanisms of closure that are effective within scientific communities may not be effective outside of them. Therefore some controversies are very extended, with lay groups continuing to press issues long after experts have arrived at some consensus (e.g. Martin 1991; Richards 1991).

As an interesting special case, scientists can also feel considerable frustration when lawyers challenge their accounts in courts. Especially in the United States, the adversarial system makes no deference to scientific expertise. (Science in the United States is also routinely deconstructed in adversarial political settings.) As a result, scientists on the stand can see their claims systematically deconstructed, and evidence that they see as solid can become suspect and irrelevant – as was seen vividly in the O. J. Simpson murder trial (Lynch 1998). The conflict between traditions has led a number of scientists to simply dismiss legal methods as not aimed at the truth and inadequate for evaluating science (e.g. Huber 1994; Koshland 1994). Attention to scientific practice complicates this picture tremendously. While there could be much done by judges and lawyers to bridge these "two cultures," the law often reacts to science in ways that are appropriate and useful for its purposes

(Jasanoff 1995; also Dreyfuss 1995). Lawyers understand, perhaps imperfectly, some of the problems with the dominant model of scientific expertise, and their work in courtrooms takes advantage of that understanding.

Like its counterpart on popularization, the dominant model of expertise assumes that scientific and technical knowledge is not tied to any context. That context-free knowledge should apply everywhere and trump all other knowledge. Yet even the most pure of pure science is tied to contexts: namely purified scientific ones. And of course there are many other ways in which science is socially constructed. When science is pulled out of its home contexts, and applied to problems in the public domain, it can fail: fail to win the trust of interested parties, fail to recognize its own social assumptions, and fail to deal adequately with messy features of real-world problems. It takes luck or hard work, of a sort that is both political and technical, to successfully apply scientific and technical expertise to public issues.

Lingering Deficits

Despite criticisms of the deficit model, there are reasons not to completely abandon it. The replacement for didactic education is presumably a number of dialogues that bring together science and publics. However, dialogue requires communication of information, which remedies shortages. Dialogue is a form of education that translates knowledge and transforms perspectives, but not without some communication of information (Davies et al. 2009).

To some extent, critics of the deficit model draw attention to other types of deficits. In particular, they draw attention to public lack of knowledge about "patronage, organization, and control" in science (Wynne 1992). Lack of scientific knowledge contributes to distrust of or unfavorable attitudes toward science, and the effects of that lack of knowledge may be compounded by lack of political knowledge about institutional decision-making (Sturgis and Allum 2004). Thus the political problem that scientists see in public knowledge deficits may remain, though it may be addressed by better education about the methods, processes, and political economies of science.

Not only scientists but also laypeople employ the deficit model. During the 2001 Foot and Mouth Disease crisis in the UK, members of a focus group argued that scientific ignorance helped to spread the disease, and also left people manipulable by the press or the government: "You can tell people anything and they will believe it, . . . if the government put out a press release telling us that it is some gerbils that is spreading it everyone would believe it" (Wright and Nerlich 2006). Laypeople's attempts to understand the crisis used the deficit model as a resource.

Critics of the deficit model may employ a romanticized vision of the lay public, resulting in an asymmetry between laypeople and experts (e.g. Durant 2008). The former are seen as reflexive, aware of different kinds of knowledge and their contexts, whereas the latter see science as universally true and applicable, trumping all other knowledge. While this might be right, the asymmetry suggests the possibility of more nuanced portrayals of scientific experts.

Finally, there are questions about expertise. To the extent that expertise is something only attributed to people, and thus a result of qualities and beliefs of the attributors rather than of the experts, we should abandon the deficit model as misleadingly attributing expertise. But to the extent that expertise is a genuine ability to know about and deal with nature, we should want experts to be final arbiters on everything in their domains (Collins and Evans 2002). We take up this issue briefly in the next chapter.

16

Expertise and Public Participation

Problems with Expertise

To the extent that scientific and technical knowledge are seen as apolitical, they are not subject to democratic contestation and oversight. The "modern consensus" divides scientific representation of nature from political representation of people (Latour 1993; also Shapin and Schaffer 1985). As we have seen throughout this book, that is an artificial division, and one that is never fully achieved in practice.

Within liberal democracies, governments can and do justify actions on the basis of scientific and technical knowledge (Ezrahi 1990). There appears to be a conflict between expertise and democracy, then, because inequalities in the distribution of knowledge and expertise undermine citizen rule (Turner 2003). Yet the past few hundred years have seen expansions of the roles of government, and of the range of governing authorized and structured by expert knowledge. It is crucial for political theory, then, to understand the nature of expertise, and decide on its legitimate scope and authority.

One approach to this issue is to turn it into an empirical question about the distribution of expertise. Genuine experts on a topic have knowledge that non-experts lack. Therefore, it would seem that if we could identify how far different forms of expertise extend we could solve the problem of political legitimacy. H. M. Collins and Robert Evans (2002) adopt this way of framing the issue, but recognize different forms of expertise held by contributors to fields, those who successfully interact with those contributors, and those who can successfully evaluate contributions. As we saw in the last chapter, members of the lay public, in addition to those with relevant credentials, often have expertise that bears on technical decisions. If we can appropriately divide spheres of expertise, then the political issue is resolved by an appropriate division of labor among different kinds of experts and also non-experts; each group can be entrusted with certain kinds of decisions.

There are a number of problems with a straightforward appeal to expertise. The solution assumes clear distinctions among different domains of expertise, but research on boundary-work, controversies, and other problems in STS suggests that distinctions will not always be straightforward. Even when it is possible to draw distinctions, there are often different levels of consensus about both technical issues and moral ones (Pielke 2007). Furthermore, an overly narrow focus on decisions ignores the different ways in which problems come into being and are framed, and the different cultures in which knowledge is used and evaluated (Wynne 2003), as well as the complexly interacting factors involved in many issues (Bijker 2001).

In response to such issues, Sheila Jasanoff (2005) describes "civic epistemologies" in Germany, the United Kingdom, and the United States that have shaped biotechnology, its institutional structures, its regulation, and public responses to it. Civic epistemologies contain a variety of related components, including such things as: styles of knowledge making; approaches to and levels of trust; practices of demonstration; accepted foundations of expertise; and assumptions about the accessibility of experts. This approach demands and identifies only local solutions to the political issue of expertise. Each culture arrives at its own civic epistemology, which becomes a locally legitimate response to the issue. That may involve deference to scientific and technical expertise, but it will be a politically generated deference.

This line of thinking complicates issues of "scientific governance" (Irwin 2008). Efforts to improve the responsiveness of science and technology to public concerns need to recognize the ways in which local values interact with many different stages in the production and application of knowledge. And in particular, efforts to improve scientific governance need to recognize that expertise and power are not two completely separate entities, but are rather outcomes of related processes (Irwin 2008).

Similar considerations affect more radical visions of alternative science. Brian Martin (2006) provides four different visions: a technocratic science for the people; a pluralistic science for the people; a science by the people, allowing popular participation; and a science shaped by a more democratic world, in which the alternative nature of science flows from the alternative nature of society. We can see these visions as each fitting with particular political situations, scientific disciplines and problems, and civic epistemologies.

Box 16.1   AIDS patient groups

What would later become "Acquired Immune Deficiency Syndrome," or "AIDS," began as "Gay Related Immune Deficiency." Over the course of the 1980s and 1990s the epidemic was devastating to American gay men, and so AIDS remained – and remains – associated with gay men. Thus the disease hit a group that *was* loosely a community, and one with some experience of activism in defense of rights. The creation of tests for the presence of antibodies to HIV meant that HIV-positive cases could be diagnosed before the onset of any symptoms; many people found themselves, their friends, family, or lovers to be living with a deadly disease, but able to devote energy to understanding that sentence and the structures around it. As a result, AIDS activists became more effective than had been earlier patient groups. Steven Epstein's (1996) study of the interaction of American AIDS activists' and AIDS researchers in the 1980s and early 1990s shows how organized non-scientists can affect not only the questions asked, but the methods used to answer them.

Activists worked within the system by lobbying government agencies and AIDS researchers. To be effective, people working "inside" had to become "lay experts" on AIDS and the research around it. They continually surprised scientists with their sophisticated understandings of the disease, the drugs being explored, the immune system, and the processes of clinical testing. Epstein describes one episode, in which a biostatistician working on AIDS trials sought out ACT UP/New York's *AIDS Treatment Research Agenda* at a conference: "I walked down to the courtyard and there was this group of guys, and they were wearing muscle shirts, with earrings and funny hair. I was almost afraid. . . ." But, reading the document, "there were many places where I found it was sensible – where I found myself saying, 'You mean, we're not doing this?' or 'We're not doing it this way?'" (quoted in Epstein 1996: 247). Third, activists took hold of parts of the research and treatment processes. Project Inform in San Francisco did its own clinical trials and epidemiological studies, and groups around the United States created buying clubs for drugs.

Among the obvious targets of activism were such things as funding for research and medical treatment. AIDS activists also demanded increased and faster access to experimental drugs. Given the then reasonable assumption that the disease invariably ended in death, patients demanded the right to decide the level of risk that they were willing to take. To avoid embarrassment, the Food and Drug Administration permitted the administration of these drugs on a "compassionate use" basis.

Perhaps more surprising was a focus on the methodology of clinical trials. After it had been shown that short-term use of AZT was effective at slowing the replication of the virus, and was not unacceptably toxic, a second phase of testing was to compare the effects of AZT and a placebo. "In blunt terms, in order to be successful the study required that a sufficient number of patients die: only by pointing to deaths in the placebo group could researchers establish that those receiving the active treatment did comparatively better" (Epstein 1996: 202). Placebos were to remain under attack, and activists sometimes successfully reshaped research to avoid the use of placebos. Also at issue were simultaneous treatments in clinical trials. Researchers expected subjects to take only the treatment that they were given, whether it was the drug or a placebo. However, patients and their advocates argued that clinical trials should mimic "real-world messiness," and should therefore allow research subjects to go about their lives, including taking alternative treatments (Epstein 1996: 257).

Public Participation in Technical Decisions

It has been a widely held assumption of, and also a result of research in, STS that more public participation in technical decision-making improves the public value and quality of science and technology (e.g. Burningham 1998; Irwin 1995). Since the 1960s both within STS and in broader contexts, there has been a push for more democratization of science and technology. This is a response against a post-World War II ideal that created an autonomous science and the independence of elite decision-makers (Lengwiler 2008).

Many of the visions of increased public participation depend on ideals of deliberative democracy. "Deliberative democracy" is a term that can stand in for an array of positions in political philosophy that encourage public participation in decision-making (e.g. Bohman 1996). These stand in opposition to positions that emphasize delegation to representative governments. Voting for representatives is a particularly inarticulate form of speech, the main touted advantage of which is its efficiency. Because of deliberative democracy's relative inefficiency, its theorists must pay particular attention to mechanisms and procedures to allow dialogues among citizens and between citizens and representatives. There are reasons to believe that procedures can be developed that allow productive dialogues to happen efficiently, even about technical issues (Hamlett 2003).

Deliberative democracy in technical decisions – about such things as nuclear waste disposal, regulation of genetically modified food, or stem cell research – should in principle lead to those decisions better reflecting public interests. Since, as we have seen, technical decisions contain normative and social assumptions, they will be improved if they have input from interested publics. The claim is not that citizens are in a better position to make decisions than experts, but that they typically have relevant knowledge and insights. It is not "that an average citizen is able to design a nuclear reactor or a river dike, but . . . more is involved in designing large projects such as nuclear power stations and water management systems than is described in the engineers' handbooks" (Bijker 2001).

But, it is difficult to show that technical decisions are improved with citizen input. The US Chemical Weapons Disposal Program provides a useful case, and a model for further study (Futrell 2003). In 1983 the US army decided that many of its chemical weapons were obsolete, and started planning their disposal. That program was accelerated in the 1990s, because of new agreements between the United States and Russia. Chemical weapons are hazardous to dismantle, and their components hazardous to dispose of. The initial approach was a "decide, announce, defend" model, that created an entirely adversarial relationship between the army and citizens at the sites selected for disposal programs. This model took enormous amounts of time, alienated the public, and produced uniform recommendations. A later participatory model, mandated by the Federal Government, appeared to be an improvement, because it opened up options. This resulted in different, and presumably more appropriate, plans for different sites (Futrell 2003).

In addition, deliberative democracy should help to legitimate decisions, because the citizen input should be intrinsically democratizing, and the processes that allow citizen input should open decisions to scrutiny. Furthermore, deliberative democracy should help to establish trust among laypeople, experts, and decision-makers, because well-constructed dialogues will allow parties to better understand each others' knowledge bases, perspectives, and concerns.

Again, the chemical weapons disposal case very clearly supports these procedural norms. Citizens say such things as, "Basically we have been designing the program, or you know, having significant input into designing the program. Instead of the Army coming in and saying 'Well this is what we are going to do,' . . . we are telling the army what we want done and the tech guys are helping us understand what is possible." And on trust,

"Ultimately this whole thing is about trust, about building trust that was destroyed over many years" (Futrell 2003).

In general, participation exercises are more successful to the extent that participants represent the population, are independent, are involved early in the decision-making process, have real influence, are engaged in a transparent process, have access to resources, have defined tasks, and engage in structured decision-making (Rowe et al., 2004, extended in Chilvers 2008).

There are a number of different kinds of public engagement mechanisms, including such things as citizens' juries, task forces, and town meetings (Rowe and Frewer 2005). The most cited model for citizen participation is the Danish consensus conference. In the 1980s, the Danish Board of Technology created the consensus conference, a panel of citizens charged with reporting and making non-binding recommendations to the Danish parliament on a specific technical topic of concern (Sclove 2000). Experts and stakeholders have opportunities to present information to the panel, but the lay group has full control over its report. While the panel is supposed to arrive at a consensus, the conference itself is typically marked by dissensus, as experts disagree, and may see their expertise on public trial (Blok 2007). The consensus conference process has been deemed a success for its ability to democratize technical decision-making without obviously sacrificing clarity and rationality, and has been extended to other parts of Europe, Japan, and the United States (Sclove 2000).

The Danish consensus conference has been successfully exported to other countries, so it has the potential to "travel well" (Einsiedel et al. 2001). However, this does not mean that every consensus conference is a success. At the first New Zealand consensus conference on biotechnology, citizen participants largely deferred to scientific and economic experts, and deferred to the Maori on their ethical concerns (Goven 2003). Although the organizers appeared to be well-meaning, they had no experience organizing such a conference and did not allow for a range of expert voices to be heard. As a result, citizen participants heard the scientific and economic experts speak with one voice, and had few resources to challenge that voice. An Australian attempt faced similar problems of insufficient organization, as well as a too-compressed time frame (Mohr 2002). A Japanese consensus conference on genetically modified crops similarly failed to raise important questions, perhaps because of the combination of the government's pre-existing commitments and the cultural importance of polite agreement in Japan (Nishizawa 2005). In general, in participants' efforts to make recommendations, consensus conferences may close down options as often as they open them up (Stirling 2008). At the very least, consensus conferences require great skill to organize, and probably require local experience and knowledge to organize well.

Citizen Science and Technology

There are alternative modes of democratizing science and technology. Society is already in science and technology. The strong programme, actor-network theory, feminist studies, and other parts of STS have all shown, at least as an approximation, the social construction of science and technology. Scientific knowledge is the result of the mobilization of resources to produce agreement among key researchers. Similarly, successful technologies are the result of the interplay among multiple actors and materials to produce artifacts that can be said to serve specific interests. Knowledge and artifacts may reflect the socialization and training of the actors who make them, and may also reflect assumptions that are more widely held. Thus if there is a problem of democracy, it is a problem of the ways in which science and technology are socially constructed, or a problem of the parts of society that participate in the constructing.

Bruno Latour, in a political manifesto, aims to bring the sciences into democracy by "blurring the distinction between nature and society durably" (2004). In the place of the modern constitution that separates nature and society, he proposes the instauration of a collective that deliberates and decides on its membership. This collective will be a Republic of things, human and nonhuman. Just as actor-network theory integrates humans and non-humans into analyses of technoscience, it might be possible to integrate humans and non-humans into technoscientific democracy. In a sense Latour's proposal is to formally recognize the centrality of science and technology to contemporary societies. But it is difficult to know how to turn that recognition into practical politics.

One route toward citizen science and technology is by making scientific and technical resources available to interested groups. In the 1970s the Netherlands pioneered the idea of "science shops," which provide technical advice to citizens, associations, and non-profit organizations (Farkas, 1999). The science shop is typically a small-scale organization that conducts scientific research in response to needs articulated by individuals or organizations lacking the resources to conduct research on their own. This idea, instantiated in many different ways, has been modestly successful, being exported to countries across Europe, and to Canada, Israel, South Africa, and the United States, though its initial popularity has waxed and waned so far (Fischer et al., 2004).

Box 16.2 Patient groups directing science

In a series of works, Michel Callon and Vololona Rabeharisoa (e.g. 2008) have displayed a form of science in which interested citizens have formed networks, organized science around their interests, and contributed to it directly.

The French association of muscular dystrophy patients (AFM) grew from a small group of patients and their parents, based on personal appeals to try to show their own humanity, and the scientific interest of their illnesses. As the AFM grew, it started participating more directly in research, not just as passive research subjects, but as active observers of their own conditions. They formed *hybrid research collectives* with scientists and scientific organizations (Callon and Rabeharisoa 2008). Eventually, owing to the phenomenal success of the AFM's fundraising Téléthon, it was able to organize and support significant amounts of scientific research. Simultaneously, it was able to use its success for political gains, improving the rights of the disabled in France, and gaining access to government services.

In its support for research, AFM made choices about how to define topics of interest. As the number of genetic diseases seemed to proliferate, it defined a number of model diseases, on the assumption that similar diseases could be treated similarly. In so doing, it reproduced some of the scientific exclusions that had led to the formation of AFM in the first place, albeit on a much smaller scale. Thus, while successful patient groups can help democratize science, expanding the range of social interests contributing to science, they do so imperfectly. And while the AFM clearly represents patient interests, some patient groups are constructed only to give the appearance of representing a social movement, but are in fact wholly supported by corporations to represent their interests: they are not grassroots organizations, but astroturf ones.

People sometimes take science into their own hands. Grassroots environmental science can be the response of people who believe that they are exposed to larger risks than are officially acknowledged. Members of communities may come together to map illnesses or measure toxins (e.g. Corburn 2005). Sick Building Syndrome, a perceived pattern of higher-than-expected illness among workers in a building, might serve as an exemplary case, for its complexity and elusiveness (Murphy 2006). As soon as a definite pollutant was identified in a building, the case was no longer one of a syndrome. One of the ways that the syndrome became visible was through consciousness raising, which had already become important to the women's health movement, and was adopted by women office workers' labor organizations. For office workers, consciousness raising encouraged a sharing of experience, both informally and through surveys, that allowed groups of workers to suggest common causes of multiple ailments. Then, through pamphlets, posters, and meetings the group experiences were systematized, validated, and communicated more widely, increasing awareness of the ailments and their possible causes. These workers were engaging in a form of popular epidemiology. Needless to say, their methods were quite different from the ones that industrial toxicologists used, making their research highly contested (Murphy 2006).

Occasionally, citizen science appears to be the most efficient way of doing research. For example, as a result of legislation and regulation in the United States, many clinical trials must be inclusive in the populations from which they draw. Thus researchers face some pressure to recruit members of minority groups. African Americans in particular tend to be unwilling to volunteer for clinical trials, because of longstanding cultural distrust of medical researchers and even physicians. Some recruiters have turned to participatory action research, in which the community takes on ownership and some organization of the trials, as a way of convincing African Americans and others to volunteer (Epstein 2007).

The broadest but most difficult route toward citizen science and technology is via a generally more egalitarian world. When groups have access to enough resources and coherent enough interests, they can direct and even participate in scientific and technological research (Box 16.2). When the resources needed to participate are relatively inexpensive and widely available, citizen science happens. We see this in the history of open-source software, where ideals of open access and opposition to large corporations have allowed coordination of contributions from hundreds and thousands of aficionados. We also see it in the histories of bicycles, musical instruments, and other popular technologies that have been influenced by many independent tinkerers (Pinch and Bijker 1987; Pinch and Bijsterveld 2004). And we also see it when specialist communities take over what they need to do their work, as when academic disciplines have created alternatives to for-profit publishing (Gunnarsdóttir 2005). These kinds of innovations can be encouraged by such things as communities providing support and validation, and discouraged by intellectual property regimes that make tinkering legally risky or expensive (von Hippel 2005). Democratized citizen science requires democratic access to resources more broadly.

Political Economies of Knowledge

As we have seen, technical knowledge is created, transferred, and even learned only through skilled work, often involving multiple actors. We might reasonably see technical knowledge as quasi-substantial, something that behaves as though it is substantial or material. It takes work and resources to create it, and moves with friction once created.

The term *knowledge economy* usually refers to an economy based on highly developed technical knowledge. It also, though, refers to economies of knowledge, structures in which knowledge is a major good, exchanged in one way or another. Clearly the first kind of knowledge economy requires the second (though not vice versa). In the new knowledge economies of the twentieth and twenty-first centuries, actors treat technical knowledge as a resource, and attempt to own or control it using mechanisms of intellectual property law. Thus business schools have contributed the discipline of knowledge management, studying how to shape the creation and flow of knowledge so that institutions can use it most efficiently and effectively.

To the extent that STS treats knowledge as quasi-substantial, we might find value in exploring the second kind of knowledge economy, even when the economies in question are non-market ones, relying on gift exchanges or efforts aimed at communal goods. Because it does not separate epistemic and political processes, STS can genuinely study scientific and technological societies, rather than treating science and technology as externalities to political processes. This last chapter is framed around the idea that among other things STS studies political economies of knowledge: the production, distribution and consumption of knowledge. Structures and circumstances vary widely, and so the discussion here can only indicate a few directions in which we might consider those political economies, suggesting further research. Here, the focus is on the commercialization of science and on STS research connected with global development. It could also have been on much more local political economies, ones instantiated in particular disciplines or specialties, with the issues that come up locally. But these have garnered much of the attention of earlier chapters.

Traditional history, philosophy, and sociology of science assume that science, at least in its ideal forms, is something like a free market of knowledge. That is, they assume that ideally all production, distribution, and consumption of knowledge is voluntary, and that it is free from regulation, interference, or fraud. Moreover, a number of people take science to largely meet that ideal (e.g. Kitcher 2001; Merton 1973; Polanyi 1962). We should not lose sight of ways in which it does meet that ideal. For example, traditional pure science is distinctive in that actors in a field are only responsive to each other, rather than to outside interests. That means that the prime audience for scientific action consists of competitors, which should create a climate of generalized skepticism.

However, as should be clear from earlier chapters, science and technology do not contain many free knowledge markets, or at least many large ones, and free markets are not in all ways desirable. Science and technology are importantly regulated by cultures and practices, by individual and institutional gatekeepers, and they are responsive to a variety of internal and external demands and forces. One way to understand the surprise that many people have when confronted with work in STS, from Thomas Kuhn's work onward, is simply that it reveals that our most successful knowledge regimes are nothing like free markets, and depend on that fact. In fact, STS's constructivism fits well with the dominant line of thinking about markets in political economy. Whereas Adam Smith, Karl Marx, and other classical and neo-classical economists believe that markets are natural outgrowths of people's desire to trade, work by historians, sociologists, political scientists, and institutional economists shows that markets have definite histories, are actively made, and are shaped by circumstances (Barma and Vogel 2008). STS would suggest the same about knowledge markets (Fuller 2007).

One of the areas in which STS has explicitly and exuberantly adopted talk of political economy is in studies of the intersection of the new commercialization of biotech and medical research, and the changes in the biotechnology and pharmaceutical industries using that research. The combination of new forms of research in these areas, their obvious commercial value, and the exploitation of research subjects makes frameworks of political economy clearly salient. Pharmaceutical companies now subcontract the bulk of their clinical research to contract research organizations or site management organizations (Mirowski and Van Horn 2005). There is also a slow but steady shifting of clinical trial activity from West to East and North to South; this "offshoring" (Petryna 2007) takes advantage of free trade in data and access to trial subjects' less expensive "experimental labor" (Cooper 2008) in less wealthy countries. Even in the US, many trial subjects are people without medical insurance, who can gain access to free medical care by enrolling in clinical trials; pharmaceutical clinical trials have become an alternative to standard medical care (Fisher 2009). Biotechnology companies need tissues and cell lines for research, and thus there is a trade in such things as embryonic stem cell lines. These "tissue economies" typically depend on gifts or donations for their initial raw materials, and depend on rhetorics of gift-giving for their legitimacy (Waldby and Mitchell 2006). Meanwhile, biotech and pharmaceutical companies are participating in a speculation based on expectations and hype around their future products. These various economies, entangled together, might constitute "biocapital," a distinctive subset of capitalist economies as a whole (Rajan 2006).

Box 17.1 Science, technology, and the formation of a modern state

On the basis of a historical study of English scientific and engineering work in Ireland in the eighteenth and nineteenth centuries, Patrick Carroll (2006) argues that science plays crucial roles in the fabric of modern states. Ireland, Carroll argues, was a "living laboratory" of English statecraft; the nineteenth-century intellectual William Nassau Sr. said, "Experiments are made in that country, and pushed to their extreme consequences." The Ordnance Survey of Ireland begun in 1825 was an extraordinarily precise and careful mapping exercise, imposing order on both the natural and human landscape and facilitating engineering projects. The land survey was integrated with a population census that gathered detailed information on England's Irish subjects, and allowed for more rational governance. Finally, Carroll explores the roles of the medical police in organizing social spaces and allowing the government to act on the health of the population; in Foucauldian terms, the state takes on the task of administering life itself, or biopower (e.g. Rabinow and Rose 2006). The state was partly constituted by its scientific and technical actions, and not just by its military and revenue-collection actions. To the extent that the Irish case is generalizable, and it is very plausible that it is, the modern state is a scientific state.

Commercialization of Research

The early 1980s appears to have been a turning point in relations between universities and industry. In the US, the passage of the Bayh–Dole Act in 1980 allowed university researchers and universities to patent research funded by Federal Government grants. The Act itself was an important point in a process of building an institutional culture of commercialization within US universities, which required that universities establish patent offices and procedures (Berman 2008). Outside the United States, a number of other countries followed suit in one way or another, attempting to harness university research to economic development via commercialization.

Although corporate research has always been intended to lead to profits, large companies once created research units that were shielded enough from business concerns so that they could pursue open-ended questions. Since the 1980s, restructuring of many companies, especially in the United States, has led to the shrinking of those laboratories, and to their changing relationships with the rest of the company. One researcher comments that "I have jumped from theoretical physics to what customers would like the most" (Varma 2000: 405). In part this is a change in the notion of the purposes of science, but it may also represent a change in people's sense of science's technological potential – as the first idea, above, of the knowledge economy suggests.

Critics might charge that the changes described are not nearly as abrupt as they are portrayed to be. But these criticisms do not take away from the feeling that dramatic changes are under way for scientific research, and that these changes are connected to the potential application of that research.

The new models of technoscientific research have been variously described. Much discussed is the idea that there is a new "mode" of knowledge production (Gibbons et al. 1994; Ziman 1994). Mode 1 was discipline-bound and problem-oriented. Mode 2 involves transdisciplinary work from a variety of types of organizations, and an application clearly in view. Some of the same authors who put forward the Mode 2 concept have developed an alternative that sees an increase in "contextualization" of science, a process in which non-scientists become involved in shaping the direction and content of specific pieces of research (Nowotny et al. 2001). There might be an increase in the importance of research that is simultaneously basic and applicable (Stokes 1997).

A second model is the academic capitalism model (Slaughter and Rhoades 2004). A combination of diminishing government revenues and a developing neoliberal ideology leads universities to look for new sources of revenue.

Telling is that at the same time that universities are promoting corporate funding of research, the capture of intellectual property, and spin-off companies, they are also creating profit centers around such things as online education and conference services, as well as signing revenue-generating contracts for food services. Sheila Slaughter and Gary Rhoades (2004) argue that there has been a change within universities from a public good regime to an academic capitalist regime; the former treated knowledge and learning as a public good and the latter treats them as resources on which institutions, faculty members, and corporations can have claims.

The academic capitalism model suffers from not paying sufficient attention to corporations and government. Another formulation of the change is in the more organizational terms of a "triple helix" of university–industry–government in interaction (Etzkowitz and Leydesdorff 1997). Governments are demanding that universities be relevant, universities are becoming entrepreneurial, and industry is buying research from universities. The change might also be put in more negative terms, that there is a crisis in key categories that support the idea of pure scientific research: the justification for and bounds of academic freedom, the public domain, and disinterestedness have all become unclear, disrupting the ethos of pure science (McSherry 2001).

Finally, we might put more weight on the sources of funding for science. Philip Mirowski and Esther Mirjam Sent (2007) describe three regimes of the organization of science, focusing on US science; their regimes correspond well with those described by in the other models. In the first regime, extending from 1890 to World War II, corporations gained rights and expanded vertically (Chandler 1977). University laboratories were patterned after corporate laboratories, and often maintained close contact with corporations. Funding for research was for corporate research, or from charitable foundations in support of projects at elite universities, or from individual researchers and universities subsidizing their own small-scale research. In the second regime, from World War II to 1980, military funding dominated research funding in the United States. Government science policy promoted democratization of the universities and basic science as a foundation for military power. Current ideals about the purity of academic science were born in this regime, in concert with and in opposition to military goals (Mirowski and Sent 2007). Corporate research laboratories continued, but as profit-making units of new conglomerates, often depending on military contracts. In the third regime, from 1980 onward, government science policy has tended to support the privatization of publicly funded research. Corporations have outsourced research and development but increased their reliance on intellectual property tools, as part of a general abandonment of the integrated firm. Academic capitalism is a response to both government and corporate changes.

Any account of the new regime of the commercialization of science needs to look beyond the United States, though. While US governments, corporations, and universities have undoubtedly led the way, and through their dominance in trade negotiations have helped to establish conditions

---

Box 17.2   Fully commercialized science: Clinical research as marketing

The number of published medical journal articles on blockbuster drugs is enormous. In recent years medical journals have published hundreds of articles per year mentioning atorvastatin (a common cholesterol-lowering drug) in the title or abstract, hundreds per year mentioning alendronate (an osteoporosis drug), and many hundreds per year mentioning fluoxetine (an antidepressant). Much less ink is spilled on drugs that sell in smaller quantities.

A major part of the reason for this huge medical interest in best-selling drugs is that many of the articles on them are parts of drug companies' "publication plans" (Sismondo 2009). Clinical research is typically performed by contract research organizations, analyzed by company statisticians, written up by independent medical writers, approved and edited by academic researchers who then serve as authors, and the whole process is organized and shepherded through to medical journals by publication planners. The research is performed at least in part because those medical journal articles advertise the drugs to physicians, a key public for pharmaceutical companies. Publication planners prefer to be involved early, so that they can map out the flow of knowledge for maximum effect. This is the *ghost management* of clinical research for purposes of marketing. The knowledge that stems from this research is created precisely for its public effects.

The best evidence suggests that 40 percent of scientific articles on major drugs are part of publication plans (Healy and Cattell 2003; Ross et al. 2008). Academic researchers, who make up the majority of the authors on these articles, play only minor roles in the research, analysis, and writing, but crucial roles in marketing: they provide legitimacy for the publications. They are also important mediators between the companies and consumers, as well as regulators (Fishman 2004). In the ghost management of medical research by pharmaceutical companies, we have a novel model of science. This is corporate science, done by many hidden workers, performed for marketing purposes, and drawing its authority from traditional academic science.

for commercialization elsewhere, it remains to be explained why commercialization has also been successful in many other parts of the developed world.

Common across the more developed world since the 1980s has been a shift in government commitment to emphasize markets and private ownership of intellectual property. International agreements on intellectual property, such as the Agreement on Trade-Related Aspects of Intellectual Property Rights, negotiated in 1994, have helped to standardize intellectual property regimes, pushing almost every nation to expand its domains of intellectual property. Changing intellectual property regimes, which have led to a "propertization" (Nowotny 2005) of science's products, have made commercialization of university research more feasible and attractive. In many countries this has been pushed further by concerted government efforts to increase national investment in research and development, asking of university science that it contribute to economic growth.

Some of the reasons may be idiosyncratic. In Canada, for example, a sudden drop in the success rate for federal grant renewals in the biomedical sciences in the mid-1980s prompted scientists to build larger labs that depended on multiple grants and other sources of funding than a single federal grant renewed for term after term (Salonius 2009). A new national science policy in the 1980s created substantial incentives for university–industry partnerships, though eventual underfunding by the Federal government guaranteed that the academic sides of those partnerships would become increasingly tied to their industry support (Atkinson-Grosjean 2006).

Thus, commercialization has been driven by some common features, like specific trade agreements, and also idiosyncratic ones, like granting structures. STS's accounts of commercialization, like its accounts of everything else, should show how this near-global phenomenon has been constructed locally.

STS and Global Development

Science and technology have at times served as important legitimators of imperialism and colonialism. When Europe started its colonial adventures the differences between European and non-European technological ability were typically small enough that they did not provide a justification for colonialism; religion served that purpose (Adas 1989; also Pyenson 1985). That changed, perhaps in part because of colonialism. By the nineteenth century, European technological development, measured in terms of military power, the mechanical amplification of human labor, or other ways, did form an explicit justification for colonialism. Technology was a symbol of Europe's modernity, and was something that Europeans could generously take to the rest of the world (Adas 1989). Often, though, that the technology transfer from Europe to its colonies resulted more in underdevelopment and dependence relationships than in the hoped-for industrialization (e.g. Headrick 1988). In broad terms, this theme has continued to be important in debates about the environmental impacts of the exportation of Western agriculture and forestry (e.g. Nandy 1988).

There is an abundance of research and writing on connections of science, technology, and development (e.g. Shrum and Shenhav 1995; Cozzens et al. 2007). This is unsurprising, given that it is widely agreed that contemporary development of nations depends heavily on science and technology. However, the constructivist tradition in STS described here has contributed modestly to this research. The field has been relatively unsuccessful at directly addressing issues of development.

Let us start with areas in which STS has made contributions. STS has been good at displaying the locality of science and technology, and the consequent conflicts and other relations between localized science and technology. Colonial and postcolonial science can be vividly marked as such, in a variety of different ways. For example, French uranium mining operations in Africa in the 1950s and '60s combined two different kinds of "rupture" talk (Hecht 2002). One was about the nuclear rupture that would provide limitless energy and restore the glory of France. The other was about the colonial rupture that would end French colonialism one colony at a time. The interplay of these two created particular hierarchies in the different national contexts.

To take another colonial example, the transplantation of the Cinchona tree (from whose bark quinine can be derived) from Peru to India, via Kew Gardens in London, was a colonial adventure performed in the name of science (Philip 1995). The East India Company's Sir Clements Markham traveled to Peru's Caravaya forest to obtain specimens, where despite fierce opposition he managed to hire local Cinchona experts to guide him to places where the trees grew, from which he took seeds and saplings. Markham portrayed himself as serving science, though he had considerably less knowledge of Cinchona than did his local informants.

To take a postcolonial example, the Giant Metrewave Radio Telescope, an Indian "big science" project, is described by its developers in terms of various features that make it the biggest of its kind (Itty 2000). But it is also connected to nation-building projects, in that its site is justified as being the best such site in the world. While nationalist rhetorics often attach themselves to big science, even in and near centers of science and technology, as in Canada (Gingras and Trépanier 1993), the rhetoric around this telescope is designed around an Indian ambivalence to modernity: India simultaneously strives for modernity but does not want to simply mimic its former colonizer (Itty 2000).

A similar relation to modernity is seen in the twentieth-century redevelopment of Ayurvedic medicine in India (Langford 2002). Ayurvedic medicine was tied to nationalist projects, which required it to be both ancient and modern. It was construed as ancient in its practices and their scientific bases. It was modern in that it was updated to incorporate aspects of allopathic medicine and also developed anew. In addition, Ayurvedic medicine focused on complaints that could be seen as cultural illnesses of colonialism, complaints that could be connected to colonial rule and the spread of Western cultural practices. Thus Ayurvedic medicine could be authentically Indian (Langford 2002).

Conflicts between knowledges can provoke a variety of responses. In Tanzania nurses (and doctors) working within Western medicine consider the traditional malady *degedege* to be malaria (Langwick 2007). There is a conflict between these nurses and traditional healers over who should treat a patient, or who should treat a patient first. Nurses in hospitals are concerned that patients treated first by healers are brought to them very late, whereas healers are concerned that patients treated first for malaria are more likely to develop *degedege*, which involves convulsions, a very hot body, and foam from the mouth. But the conflict is not an all-or-nothing one. In this context, bodies, diseases, and treatments are articulated in relation to both traditions, as nurses and healers attempt to find stable diagnoses and explanations that will bring patients to them at the right times (Langwick 2007). When thinking normatively about relations between Western and non-Western knowledges we should also be open to a variety of possibilities (Harding 2006).

Moreover, conflicts and other relationships between traditional and Western medicines are not just seen in the homes of those traditional medicines. There is a huge demand for traditional remedies in the United States and Europe, where consumers long for the "magic" of those remedies (Adams 2002). The magic, though, is very difficult to stabilize. Tibetan doctors see their remedies as scientific, and are very interested in scientific tests of their efficacy. Those scientific tests, though, are constructed in ways that make it very difficult for the remedies to pass, because all Tibetan diagnostic and disease categories are first translated into Western ones, contexts of treatment are ignored and replaced by contexts of scientific testing, and medicines are reduced to ingredients. "Discerning between the scientific and the magical/spiritual is itself a biopolitical function of the modern state" (Adams 2002). That discernment, though, means if some medicine should happen to be successful, credit and profits will go to a pharmaceutical company developer. Thus, some pharmaceutical companies are interested in transferring Tibetan remedies to North America and Europe, but in so doing they leave behind almost everything that is specifically Tibetan. Meanwhile, remedies that have not had their magical properties exorcised are made illegal or allowed only as "supplements," no matter how much Western consumers might be interested in the magic.

Bioprospecting is a point of contact between science and development that has attracted considerable attention, mostly critical. Bioprospecting is the search for biologically active molecules with commercial value, typically in pharmaceutical or agricultural uses. Bioprospectors take advantage of local knowledge of plants and animals, to find substances that have better-than-average chances of being valuable. To avoid being seen as *biopiracy* (Shiva 1997), bioprospecting has increasingly made use of an ethical and legal framework of benefit-sharing, under which communities from whom knowledge and substances are taken are promised benefits from any commercial uses of those substances. The benefits are modest in theory, and are even more so in practice. Nonetheless, the benefit agreements themselves create political problems, because the boundaries of communities holding local ethnobiological knowledge have to be decided for legal purposes, but more importantly those agreements have the consequence of taking objects from the public domain and subjecting them to regimes of intellectual property (Escobar 1999; Hayden 2003).

Current intellectual property regimes create a number of development issues. In the nineteenth century it was acceptable for nations not to have patent systems, or to have systems that were weak enough that local entrepreneurs could freely reverse engineer, copy, and sell foreign technologies. This was an important strategy of technological development, and developing countries like the Netherlands and Switzerland took advantage of it. With increasing standardization and integration of patent systems, from the Paris Convention of 1883, the first international patent treaty, to the current day, the situation has changed. Current patent agreements lock in inequalities (Carolan 2008). Developing nations find it difficult to opt out of international patent regimes, both because of general pressure from wealthy countries and because of focused pressure from patent holders. For example, in the period from 2004 to 2006, Monsanto used its European patent on Roundup Ready soybeans to apply pressure to Argentina, which was not collecting royalties on soybeans containing the Roundup Ready gene, and was exporting those soybeans to Europe. Monsanto sued European

Box 17.3   The fluidity of appropriate technologies

Given that knowledges and technologies move only with difficulty, it is particularly difficult to move technologies from more developed countries to less developed ones. For example, a French photoelectric lighting kit was limited by its origins in a technologically rich and orderly environment, not reproduced in its intended environments in Africa (Akrich 1992). An Australian dairy could not be reproduced in East Timor, and even a translated form of it dealt poorly with local natural and cultural conditions (Shepherd and Gibbs 2006).

In an account of a water pump, the Zimbabwe Bush Pump "B" Type, Marianne de Laet and Annemarie Mol (2000) explore what makes for an *appropriate technology*. On their account, it is the *fluidity* of the Bush Pump that makes it appropriate for use in Zimbabwean villages. The Bush Pump is simple, durable, and resilient, but has undefined boundaries and is flexible in its design. It is a source of fresh water, a promoter of health, and a part of nation-building.

The Bush Pump can withstand losing many of the bolts that hold it together. It can be repaired by villagers with a few simple tools and parts. Or, with more ingenuity it can be jury rigged and even redesigned by villagers working with what they have available. Villagers install the pumps themselves, on precise but simple instructions, and using standardized drilling equipment. Without the participation of a number of members of the community, a pump will not work; and so de Laet and Mol argue that the community itself is part of the infrastructure of the pump.

The Bush Pump succeeds in part because it is in the public domain. Its most clear "inventor" denies both his inventor status and ownership of the pump, and as a result it belongs to the Zimbabwean people, who not only do not have to pay royalties to use it but can proudly claim ownership of it. As a Zimbabwean technology, each installation of a pump helps to build the nation in both concrete and symbolic ways; pump installation is an action not only of the village but of the nation, a more dispersed version of the technological state action described by Carroll (Box 17.1).

importers for high royalties on the beans, those importers sought other sources, and eventually Argentine producers and the Argentine government agreed to pay somewhat lower royalties. This effectively extended Monsanto's patent and enforcement of it into a country with a different patent system (Kranakis 2007).

Cozzens et al. (2007) argue that STS's success at showing conflicts has stood in the way of its offering positive lessons for development. Meanwhile, they argue that the disciplines of economics and political science have likewise been unsuccessful, because they have focused on growth and innovation respectively, rather than on freedom – Amartya Sen (2000) describes development as "a process expanding the real freedoms that people enjoy." Economic growth and industrial innovation can contribute to that, but need not be the only forms that development takes. In fact, it is well known that inequality impedes even simple economic growth. And on Sen's understanding of development, technological innovation plays little role, because most innovation is aimed at the wealthy markets of already well-developed parts of the world.

The failure of the linear model of innovation (Chapter 9) suggests that investment in science will not necessarily lead to economic growth. That is borne out by empirical evidence that in less developed countries there is no relationship between levels of (Western) scientific knowledge and economic performance (Shenhav and Kamens 1991), even while there is a strong relationship in more developed countries. Economists, too, have objected to the linear model. The very popular National Innovation Systems concept (NIS) is a framework for describing the institutions – specific state agencies, corporations, and academic institutions – and their linkages that contribute to technological growth. NIS, which arose in part as a rejection of what little neoclassical economics had to say on the issue of technology, understands growth as a result of an efficient creation and transfer of knowledge, skills, and artifacts (Sharif 2006). NIS by itself, though, does not provide a formula for technological growth, and instead, like STS, points to the need for local case studies.

STS should start to make its contribution to development studies by applying, to the extent that it can, its established results to contexts of development. What follows is an extremely schematic research agenda for STS, suggesting some themes from earlier chapters that should be applied to empirical studies of science, technology, and development.

STS examines how things are constructed. In so doing, the field has largely adopted an agent- or action-centered perspective, rather than a structure-centered one: STS's constructions are active. Much of the success of the field has stemmed from an insistent localism and materialism, seeing macro-level structures as constituted by and having their effects in micro-level actions. Thus STS's contributions to development studies will likely take the form of case studies of the successes and failures to link science, technology, and increased freedom. Given some of the consistent patterns of underdevelopment, it might seem that STS's localism and action-centered perspectives are less appropriate when applied to issues of development. But the same has been said of scientific and technological change in developed contexts, and that has not prevented action-centered STS from flourishing. Moreover, in a review of the broader literature on science, technology, and development, Wesley Shrum and Yehouda Shenhav (1995) see a striking agreement that "science and technology should be viewed in terms of context-specific forms of knowledge and practice that interact with a set of globally distributed social interests."

As conflicts among knowledges and forms of knowledge show, a key problem for studies of global science is understanding relationships between science in less developed and in more developed countries, between science in centers and peripheries (Moravcsik and Ziman 1975). In interactions between people in peripheral and central locations typical models privilege those whose expertise is validated by central networks – just as do typical models of interactions between laypeople and scientists. Despite sizeable numbers of scientists, the less developed world remains scientifically peripheral; it is "dependent" science: "What is considered scientific knowledge in a dependent context is only that which has been made legitimate in the centre. It is then imitated in the periphery through the operation of pervasive dependent social and cultural mechanisms . . . The fundamental and the basic core knowledge grows largely in the West and is transferred to developing countries in the context of a dependent intellectual relationship" (Goonatilake 1993: 260) Interestingly, that insight also applies to peripheral locations within the center, since people outside the core set of researchers in a field have little influence on the field's intellectual shape. Once again we are back to questions about the mechanisms of stratification (Chapter 4).

Stratification in science can be partially explained by the cumulative advantage hypothesis, that large differences in status can be the result of the accumulation of many small differences. Somebody who faces even small levels of discrimination along a career path will eventually find it much more difficult to progress along that path. Structures that discriminate can have their effects in concrete interactions, sometimes trivial-seeming ones. In the context of development, researchers could be alert to ways in which people at peripheries of the scientific and high-technological worlds can build cumulative advantage, rather than disadvantage. And aiming at development as freedom, researchers might particularly focus on issues of equity within societies. Gender, for example, may have substantial effects on the options available to people, and negative effects may be accentuated where networks are already limited (Campion and Shrum 2004), perhaps by traditional cultures (Gupta 2007).

From the strong programme (Chapter 5) STS has acquired an appreciation of the methodological importance of symmetry. Beliefs considered true or rational on the one hand, or false or irrational on the other, should be explained using the same resources. In the context of development, we might extend the principle of symmetry to explain increases and decreases of freedom using the same resources. The same types of actors who can combine to promote growth can also combine to retard it. This might suggest a symmetrizing appropriation of something like the National Innovation Systems concept mentioned above.

As we saw especially in the context of actor-network theory (Chapter 8), what appear to be independent pieces of science and technology are always parts of broader networks, and depend for their working on those networks. When they become well established we can see those networks as systems, with the entrenchment of systems. A lone claim or object will do nothing if it is not supported by other ones, and by the other actors who bring them to bear on each other. This is part of the enormous difficulty of the development project. One cannot simply create lone facts or artifacts without networks to support them, yet local networks may be weak and international ones may be unhelpful or even stand in opposition. And for organizations international networks might have to have a kind of organic integrity, or risk destabilizing the local ones (Shrum 2000).

A technological frame is the set of practices and the material and social infrastructure built up around an artifact (Chapter 9). It guides the meanings given to an artifact, and through that the ways in which it is used and the directions in which it might be developed. Yet at the same time, STS shows that there is interpretive flexibility around technologies. Technological frames are pliant. Thus when STS follows technologies in less developed countries it can be attentive to the established meanings, both local and non-local, that enable and limit those technologies.

Tacit knowledge is that knowledge that cannot be easily formalized (Chapter 10). Typically, tacit knowledge can be most easily communicated through a socialization process, in which the learner enters into parts of the culture of the teacher. In those cases in which development is a project of duplicating in less developed countries objects and systems found in more developed ones, moving tacit knowledge will be a central issue and problem.

The power of science and technology must stem at least in part from abilities to establish formal objectivity, to black box some claims and objects, to treat material in rule-bound ways (Chapter 12). Otherwise, every scientific or technical action would be new, unable to apply or build on previous work. Yet, as we saw, human action is shot through and through with interpretation. This apparent conflict is resolved when we understand that interpretation can be in the aid of objectivity, and not merely an obstacle to it. In the context of development, STS could study cases in which interpretation does and does not support objectivity, and does and does not support the establishment of stable networks.

STS has developed a variety of perspectives on the democratic control of science and technology (Chapter 16). With the goal of increasing freedoms in mind, those perspectives are surely as or more "important" in the context of development as they are in the contexts in which they have been created. Thus STS can contribute to development studies a set of ways of thinking about democratic control of technical issues. For example, recognizing that there are local civic epistemologies that differ even among countries in Western Europe and North America should allow for the recognition and nurturing of legitimate civic epistemologies elsewhere.

The constructivist view brings to the fore the complexity of real-world science and technology – and therefore can contribute to their success. Successful science and technology in the public sphere, as in contexts of global development, can be the result of the co-construction of science and politics. Successful public science can be the result of careful adjustment of scientific knowledge to make it fit new contexts; assumptions that restrict science to purified domains have to be relaxed, and work has to be done to attune scientific knowledge to the knowledges of others. An exactly analogous argument can be made about technology.

Scientific and technical expertise is seen as universally applicable, because scientific and technical knowledge is universal. As we have seen, there are senses in which scientific and technical knowledge aims at universality. Formal objectivity – the denial of subjectivity – creates a kind of universality. Objective procedures are ones that, if correctly followed, will produce the same results given the same starting points. The highly abstract character of much scientific and technical knowledge also gives it a kind of universality. Abstractions are valued precisely because they leave aside messy concrete details.

Yet, to the extent that abstract scientific and technical claims describe the real world, they do not describe *particular* features of it. Thus the senses in which scientific knowledge and technological artifacts are universal can also be seen as a kind of locality and particularity. Objectivity and abstraction shape science and technology to particular contexts – the insides of laboratories, highly constructed environments, and kinds of Platonic worlds – albeit contexts interesting in their reproducibility or their lack of concrete location. While these interesting shapings allow science to be widely applicable, they also limit its applicability in concrete situations, such as ones demanded by projects of development. In addition, scientific and technical knowledge is local and particular in that it is the result of local and particular processes, the result of network-building in the context of disciplinary and other cultural forces, the result of rhetorical and political work in those same contexts, and sometimes the outcome of controversies whose outcomes appear contingent. If it is to be made to work at its best, scientific and technical knowledge needs to be seen as situated in social and material spaces. It is, in short, constructed.

References

Abrahaam, Itty (2000) "Postcolonial Science, Big Science, and Landscape." In R. Reid and S. Traweek, eds, *Doing Science + Culture*. New York: Routledge, 49–70.

Ackermann, Robert J. (1985) *Data, Instruments and Theory: A Dialectical Approach to Understanding Science*. Princeton: Princeton University Press.

Adams, Vincanne (2002) "Randomized Controlled Crime: Postcolonial Sciences in Alternative Medicine Research." *Social Studies of Science* 32: 659–90.

Adas, Michael (1989) *Machines as the Measures of Men: Science, Technology, and Ideologies of Western Dominance*. Ithaca, NY: Cornell University Press.

Akrich, Madeleine (1992) "The De-Scription of Technical Objects." In W. Bijker and J. Law, eds, *Shaping Technology – Building Society: Studies in Sociotechnical Change*. Cambridge, MA: The MIT Press, 205–44.

Alder, Ken (1998) "Making Things the Same: Representation, Tolerance and the End of the Ancien Régime in France." *Social Studies of Science* 28: 499–546.

Allison, Paul D. and J. Scott Long (1990) "Departmental Effects on Scientific Productivity." *American Sociological Review* 55: 469–78.

Amann, K. and K. Knorr Cetina (1990) "The Fixation of (visual) Evidence." In M. Lynch and S. Woolgar, eds, *Representation in Scientific Practice*. Cambridge, MA: The MIT Press, 85–122.

Appel, Toby (1994) "Physiology in American Women's Colleges: The Rise and Decline of a Female Subculture." *Isis* 85: 26–56.

Ashmore, Malcolm (1989) *The Reflexive Thesis: Wrighting Sociology of Scientific Knowledge*. Chicago: University of Chicago Press.

Ashmore, Malcolm, Michael Mulkay, and Trevor Pinch (1989) *Health and Efficiency: A Sociology of Health Economics*. Milton Keynes: Open University Press.

Atkinson-Grosjean, Janet (2006) *Public Science, Private Interests: Culture and Commerce in Canada's Networks of Centres of Excellence*. Toronto: University of Toronto Press.

Ayer, A. J. (1952) *Language, Truth and Logic*, 2nd edn (first published 1936). New York: Dover.

Bachelard, Gaston (1984) *The New Scientific Spirit*. Tr. A. Goldhammer (first published 1934). Boston: Beacon Press.

Baker, G. P. and P. M. S. Hacker (1984) *Scepticism, Rules, and Language*. Oxford: Blackwell.

Barad, Karen (2007) *Meeting the Universe Halfway: Quantum Physics and the Entanglement of Matter and Meaning*. Durham, NC: Duke University Press.

Barbercheck, Mary (2001) "Mixed Messages: Men and Women in Advertisements in Science." In M. Wyer, M. Barbercheck, D. Giesman, H. Ö. Öztürk, and M. Wayne, eds, *Women, Science, and Technology: A Reader in Feminist Science Studies*. New York: Routledge, 117–31.

Barma, Naazneen H. and Steven K. Vogel (2008) *The Political Economy Reader: Markets as Institutions*. New York: Routledge.

Barnes, Barry (1982) *T. S. Kuhn and Social Science*. New York: Columbia University Press.

Barnes, Barry and David Bloor (1982) "Relativism, Rationalism and the Sociology of Knowledge." In M. Hollis and S. Lukes, eds, *Rationality and Relativism*. Oxford: Oxford University Press, 21–47.

Barnes, Barry, David Bloor, and John Henry (1996) *Scientific Knowledge: A Sociological Analysis*. Chicago: University of Chicago Press.

Barnes, S. B. and R. G. A. Dolby (1970) "The Scientific Ethos: A Deviant Viewpoint." *Archives of European Sociology* 11: 3–25.

Bazerman, Charles (1988) *Shaping Written Knowledge: The Genre and Activity of the Experimental Article in Science*. Madison: University of Wisconsin Press.

Ben-David, Joseph (1991) *Scientific Growth: Essays on the Social Organization and Ethos of Science*. Berkeley: University of California Press.

Benbow, Camilla P. and Julian C. Stanley (1980) "Sex Differences in Mathematical Ability: Factor or Artifact?" *Science* 220: 1262–64.

Berg, Marc (1997) "Of Forms, Containers, and the Electronic Medical Record: Some Tools for a Sociology of the Formal." *Science, Technology, & Human Values* 22: 403–33.

Berger, Peter L. and Thomas Luckmann (1966) *The Social Construction of Reality: A Treatise in the Sociology of Knowledge*. Garden City, NY: Doubleday.

Berman, Elizabeth Popp (2008) "Why Did Universities Start Patenting? Institution-building and the Road to the Bayh–Dole Act." *Social Studies of Science* 38: 835–71.

Biagioli, Mario (1993) *Galileo Courtier: The Practice of Science in the Culture of Absolutism*. Chicago: University of Chicago Press.

Biagioli, Mario (1995) "Knowledge, Freedom, and Brotherly Love: Homosociality and the Accademia dei Lincei." *Configurations* 2: 139–66.

Bijker, Wiebe E. (1995) *Of Bicycles, Bakelites, and Bulbs: Toward a Theory of Sociotechnical Change*. Cambridge, MA: MIT Press.

Bijker, Wiebe E. (2001) "Understanding Technological Culture through a Constructivist View of Science, Technology, and Society." In Stephen H. Cutliffe and Carl Mitcham, eds, *Visions of STS: Counterpoints in Science, Technology, and Society Studies*. Albany, NY: SUNY Press.

Bimber, Bruce (1994) "Three Faces of Technological Determinism." In M. R. Smith and L. Marx, eds, *Does Technology Drive History? The Dilemmas of Technological Determinism.* Cambridge, MA: MIT Press, 79–100.

Biology and Gender Study Group (1989) "The Importance of Feminist Critique for Contemporary Cell Biology." In N. Tuana, ed., *Feminism and Science.* Bloomington: Indiana University Press, 172–87.

Bird, Alexander (2000) *Thomas Kuhn.* Princeton: Princeton University Press.

Bleier, Ruth (1984) *Science and Gender: A Critique of Biology and Its Theories on Women.* New York: Pergamon Press.

Blok, Anders (2007) "Experts on Public Trial: On Democratizing Expertise through a Danish Consensus Conference." *Science and Public Policy* 16: 163–82.

Bloor, David (1978) "Polyhedra and the Abominations of Leviticus." *British Journal for the History of Science* 11: 245–72.

Bloor, David (1991) *Knowledge and Social Imagery,* 2nd edn (first published 1976). Chicago: University of Chicago Press.

Bloor, David (1992) "Left and Right Wittgensteinians." In A. Pickering, ed., *Science as Practice and Culture.* Chicago: University of Chicago Press, 266–82.

Bogen, Jim and Jim Woodward (1988) "Saving the Phenomena." *Philosophical Review* 97: 303–52.

Bogen, Jim and Jim Woodward (1992) "Observations, Theories, and the Evolution of the Human Spirit." *Philosophy of Science* 59: 590–611.

Bohman, James (1996) *Public Deliberation: Pluralism, Complexity, and Democracy.* Cambridge, MA: MIT Press.

Booth, Wayne C. (1961) *Rhetoric of Fiction.* Chicago: University of Chicago Press.

Boumans, Marcel (1999) "Built in Justification." In Mary S. Morgan and Margaret Morrison, eds, *Models as Mediators: Perspectives on Natural and Social Science.* Cambridge: Cambridge University Press, 66–96.

Bourdieu, Pierre (1999) "The Specificity of the Scientific Field and the Social Conditions of the Progress of Reason." Tr. Richard Nice. In M. Biagioli, ed., *The Science Studies Reader* (first published 1973). New York: Routledge, 31–50.

Bowker, Geoffrey C. (1994) *Science on the Run: Information Management and Industrial Geophysics at Schlumberger, 1920–1940.* Cambridge, MA: The MIT Press.

Bowker, Geoffrey C. and Susan Leigh Star (2000) *Sorting Things Out: Classification and it's Consequences,* 1st edn. Cambridge, MA: The MIT Press.

Boyd, Richard (1979) "Metaphor and Theory Change: What is 'Metaphor' a Metaphor For?" In A. Ortony, ed., *Metaphor and Thought.* Cambridge: Cambridge University Press, 356–408.

Boyd, Richard N. (1984) "The Current Status of Scientific Realism." In J. Leplin, ed., *Scientific Realism.* Berkeley: University of California Press, 41–82.

Boyd, Richard N. (1985) "Lex Orandi est Lex Credendi." In P. Churchland and C. Hooker, eds, *Images of Science.* Chicago: Chicago University Press, 3–34.

Boykoff, Maxwell T. and Jules M. Boykoff (2004) "Balance as Bias: Global Warming and the U.S. Prestige Press." *Global Environmental Change* 14(2): 125–36.

Brannigan, Augustine (1981) *The Social Basis of Scientific Discoveries.* Cambridge: Cambridge University Press.

Brante, Thomas and Margareta Hallberg (1991) "Brain or Heart? The Controversy over the Concept of Death." *Social Studies of Science* 21: 389–413.

Breslau, Daniel and Yuval Yonay (1999) "Beyond Metaphor: Mathematical Models in Economics as Empirical Research." *Science in Context* 12: 317–32.

Brewer, William F. and Bruce L. Lambert (2001) "The Theory-Ladenness of Observation and the Theory-Ladenness of the Rest of the Scientific Process." *Philosophy of Science* 68 (Proceedings): S176–86.

Brighton Women & Science Group (1980) *Alice through the Microscope: The Power of Science over Women's Lives.* London: Virago.

Broad, William and Nicholas Wade (1982) *Betrayers of the Truth: Fraud and Deceit in the Halls of Science.* New York: Simon and Schuster.

Brown, James Robert (2001) *Who Rules in Science? An Opinionated Guide to the Wars.* Cambridge, MA: Harvard University Press.

Bucchi, Massimiano (1998) *Science and the Media.* London: Routledge.

Bucciarelli, Louis L. (1994) *Designing Engineers.* Cambridge, MA: MIT Press.

Burningham, Kate (1998) "A Noisy Road or Noisy Resident? A Demonstration of the Utility of Social Constructivism for Analysing Environmental Problems." *Sociological Review* 46: 536–63.

Burningham, Kate and Geoff Cooper (1999) "Being Constructive: Social Constructivism and the Environment." *Sociology* 33: 297–316.

Butterfield, Herbert (1931) *The Whig Interpretation of History.* London: Bell.

Callon, Michel (1986) "Some Elements of a Sociology of Translation: Domestication of the Scallops and the Fishermen of St. Brieuc Bay." In J. Law, ed., *Power, Action and Belief.* London: Routledge & Kegan Paul, 196–233.

Callon, Michel (1987) "Society in the Making: The Study of Technology as a Tool for Sociological Analysis." In W. E. Bijker, T. P. Hughes, and T. J. Pinch, eds, *The Social Construction of Technological Systems: New Directions in the Sociology and History of Technology.* Cambridge, MA: MIT Press, 83–103.

Callon, Michel and Bruno Latour (1992) "Don't Throw the Baby Out with the Bath School! A Reply to Collins and Yearley." In A. Pickering, ed., *Science as Practice and Culture.* Chicago: University of Chicago Press, 343–68.

Callon, Michel and John Law (1989) "On the Construction of Sociotechnical Networks: Content and Context Revisited." In R. A. Jones, L. Hargens, and A. Pickering, eds, *Knowledge and Society, Vol. 8: Studies in the Sociology of Science Past and Present.* Greenwich, CT: JAI Press, 57–83.

Callon, Michel and John Law (1995) "Agency and the Hybrid *Collectif.*" *South Atlantic Quarterly* 94: 481–507.

Callon, Michel and Vololona Rabeharisoa (2008) "The Growing Engagement of Emergent Concerned Groups in Political and Economic Life: Lessons from the French Association of Neuromuscular Disease Patients." *Science, Technology, & Human Values* 33: 230–61.

Calvert, Jane (2006) "What's Special about Basic Research?" *Science, Technology & Human Values* 31: 199–220.

Cambrosio, Alberto, Peter Keating, and Michael Mackenzie (1990) "Scientific Practice in the Courtroom: The Construction of Sociotechnical Identities in a Biotechnology Patent Dispute." *Social Problems* 37: 275–93.

Cambrosio, Alberto, Camille Limoges, and Denyse Pronovost (1990) "Representing Biotechnology: An Ethnography of Quebec Science Policy." *Social Studies of Science* 20: 195–227.

Campion, Patricia and Wesley Shrum (2004) "Gender and Science in Development: Women Scientists in Ghana, Kenya, and India." *Science, Technology, & Human Values* 29: 459–85.

Cantor, G. N. (1975) "A Critique of Shapin's Social Interpretation of the Edinburgh Phrenology Debate." *Annals of Science* 32: 245–56.

Carnap, Rudolf (1952) *The Logical Structure of the World*. Tr. R. George (first published 1928). Berkeley: University of California Press.

Carolan, Michael S. (2008) "Making Patents and Intellectual Property Work: The Asymmetrical 'Harmonization' of TRIPS." *Organization and Environment* 21: 295–310.

Carroll, Patrick (2006) *Science, Culture, and Modern State Formation*. Berkeley: University of California Press.

Cartwright, Nancy (1983) *How the Laws of Physics Lie*. Oxford: Oxford University Press.

Case, Donald O. and Georgeann M. Higgins (2000) "How Can We Investigate Citation Behavior? A Study of Reasons for Citing Literature in Communication." *Journal of the American Society for Information Science* 51: 635–45.

Casper, Monica J. and Adele E. Clarke (1998) "Making the Pap Smear into the 'Right Tool' for the Job: Cervical Cancer Screening in the USA, circa 1940–95." *Social Studies of Science* 28: 255–90.

Chandler, Alfred D., Jr. (1977) *The Visible Hand: The Managerial Revolution in American Business*. Cambridge, MA: Harvard University Press.

Charlesworth, Max, Lyndsay Farrall, Terry Stokes, and David Turnbull (1989) *Life Among the Scientists: An Anthropological Study of an Australian Scientific Community*. Oxford: Oxford University Press.

Chilvers, Jason (2008) "Deliberating Competence: Theoretical and Practitioner Perspectives on Effective Participatory Appraisal Practice." *Science, Technology, & Human Values* 33(2): 155–85.

Churchland, Paul (1988) "Perceptual Plasticity and Theoretical Neutrality: A Reply to Jerry Fodor." *Philosophy of Science* 55: 167–187.

Clark, Tim and Ron Westrum (1987) "Paradigms and Ferrets." *Social Studies of Science* 17: 3–3e.3.

Clarke, Adele E. (1990) "A Social Worlds Research Adventure: The Case of Reproductive Science." In S. E. Cozzens and T. F. Gieryn, eds, *Theories of Science in Society*. Bloomington: Indiana University Press, 15–42.

Clause, Bonnie Tocher (1993) "The Wistar Rat as a Right Choice: Establishing Mammalian Standards and the Ideal of a Standardized Mammal." *Journal of the History of Biology* 26: 329–49.

Cockburn, Cynthia (1983) *Brothers: Male Dominance and Technological Change*. London: Pluto Press.

Cockburn, Cynthia (1985) *Machinery of Dominance: Women, Men and Technical Know-how*. London: Pluto Press.

Cole, Jonathan R. (1981) "Women in Science." *American Scientist* 69: 385–91.

Cole, Jonathan R. and Stephen Cole (1973) *Social Stratification in Science*, 1st edn. Chicago and London: The University of Chicago Press.

Cole, Jonathan R. and Burton Singer (1991) "A Theory of Limited Differences: Explaining the Productivity Puzzle in Science." In H. Zuckerman, J. R. Cole, and J. T. Bruer, eds, *The Outer Circle: Women in the Scientific Community*. New York: W. W. Norton, 188–204.

Cole, Simon A. (1996) "Which Came First, the Fossil or the Fuel?" *Social Studies of Science* 26: 733–66.

Cole, Simon A. (2002) *Suspect Identities: A History of Fingerprinting and Criminal Identification*. Cambridge, MA: Harvard University Press.

Collins, H. M. (1974) "The TEA Set: Tacit Knowledge and Scientific Networks." *Science Studies* 4: 165–86.

Collins, H. M. (1990) *Artificial Experts: Social Knowledge and Intelligent Machines*. Cambridge, MA: MIT Press.

Collins, H. M. (1991) *Changing Order: Replication and Induction in Scientific Practice*, 2nd edn (1st edn 1985). Chicago: Chicago University Press.

Collins, H. M. (1996) "In Praise of Futile Gestures: How Scientific is the Sociology of Scientific Knowledge?" *Social Studies of Science* 26: 229–44.

Collins, H. M. (1999) "Tantalus and the Aliens: Publications, Audiences, and the Search for Gravitational Waves." *Social Studies of Science* 19: 163–97.

Collins, H. M. and Robert Evans (2002) "The Third Wave of Science Studies: Studies of Expertise and Experience." *Social Studies of Science* 32: 235–96.

Collins, H. M. and T. J. Pinch (1982) *Frames of Meaning: The Social Construction of Extraordinary Science*. London: Routledge & Kegan Paul.

Collins, H. M. and S. Yearley (1992) "Epistemological Chicken." In A. Pickering, ed., *Science as Practice and Culture*. Chicago: University of Chicago Press, 301–26.

Collins, Harry and Martin Kusch (1998) *The Shape of Actions: What Humans and Machines Can Do*. Cambridge, MA: MIT Press.

Collins, Harry and Trevor Pinch (1993) *The Golem: What Everyone Should Know about Science*. Cambridge: Cambridge University Press.

Collins, Randall (1998) *The Sociology of Philosophies: A Global Theory of Intellectual Change*. Cambridge, MA: Harvard University Press.

Conley, Thomas M. (1990) *Rhetoric in the European Tradition*. Chicago: University of Chicago Press.

Constant II, Edward W. (1984) "Communities and Hierarchies: Structure in the Practice of Science and Technology." In R. Laudan, ed., *The Nature of Technological Knowledge*. Dordrecht: D. Reidel, 27–46.

Cooper, Melinda (2008) "Experimental Labour – Offshoring Clinical Trials to China." *East Asian Science, Technology and Society* 2: 73–92.

Corburn, Jason (2005) *Street Science: Community Knowledge and Environmental Health Justice*. Cambridge, MA: MIT Press.

Cowan, Ruth Schwartz (1983) *More Work for Mother: The Ironies of Household Technology from the Open Hearth to the Microwave*. New York: Basic Books.

Cozzens, Susan E., Sonia Gatchair, Kyung-Sup Kim, Gonzalo Ordóñez, and Anupit Supnithadnaporn (2007) "Knowledge and Development." In E. J. Hackett, O. Amsterdamska, M. Lynch and J. Wajcman, eds, *The Handbook of Science and Technology Studies*, 3rd edn. Cambridge, MA: MIT Press, 787–811.

Crist, Eileen (2004) "Against the Social Construction of Nature and Wilderness." *Environmental Ethics* 26: 5–23.

Crowe, Michael J. (1988) "Ten Misconceptions about Mathematics and Its History." In W. Aspray and P. Kitcher, eds, *History and Philosophy of Modern Mathematics*. Minneapolis: University of Minnesota Press, 260–77.

Cummiskey, D. (1992) "Reference Failure and Scientific Realism: A Response to the Meta-induction." *British Journal for the Philosophy of Science* 43: 21–40.

Cutliffe, Stephen H. (2000) *Ideas, Machines, and Values: An Introduction to Science, Technology, and Society Studies*. Lanham, MD: Rowman & Littlefield.

Daston, Lorraine (1991) "Baconian Facts, Academic Civility, and the Prehistory of Objectivity." *Annals of Scholarship* 8: 337–63.

Daston, Lorraine (1995) "The Moral Economy of Science." *Osiris* 10: 3–24.

Daston, Lorraine (2008) "On Scientific Observation." *Isis* 99: 97–110.

Daston, Lorraine and Peter Galison (1992) "The Image of Objectivity." *Representations* 40: 83–128.

Davidson, Donald (1974) "On the Very Idea of a Conceptual Scheme." *Proceedings of the American Philosophical Association* 47: 5–20.

Davies, Sarah, Ellen McCallie, Elin Simonsson, Jane L. Lehr, and Sally Duensing (2009) "Discussing Dialogue: Perspectives on the Value of Science Dialogue Events that do not Inform Policy." *Public Understanding of Science* 18: 338–53.

Davis, Natalie Zemon (1995) *Women on the Margins: Three Seventeenth-Century Lives*. Cambridge, MA: Harvard University Press.

Dawkins, Richard (1976) *The Selfish Gene*. Oxford: Oxford University Press.

De Laet, Marianne and Annemarie Mol (2000) "The Zimbabwe Bush Pump: Mechanics of a Fluid Technology." *Social Studies of Science* 30: 225–63.

De Solla Price, D. J. (1986) *Little Science, Big Science and Beyond* (first published 1963). New York: Columbia University Press.

Dear, Peter (1995a) "Cultural History of Science: An Overview with Reflections." *Science, Technology & Human Values* 20: 150–70.

Dear, Peter (1995b) *Discipline and Experience: The Mathematical Way in the Scientific Revolution*. Chicago: University of Chicago Press.

Delamont, Sara (1989) *Knowledgeable Women: Structuralism and the Reproduction of Elites*. London: Routledge.

Delamont, Sara and Paul Atkinson (2001) "Doctoring Uncertainty: Mastering Craft Knowledge." *Social Studies of Science* 31: 87–107.

Delborne, Jason A. (2008) "Transgenes and Transgressions: Scientific Dissent as Heterogeneous Practice." *Social Studies of Science* 38: 509–41.

Demeritt, David (2001) "The Construction of Global Warming and the Politics of Science." *Annals of the Association of American Geographers* 91: 307–37.

Dewey, John (1929) *The Quest for Certainty: A Study of the Relation of Knowledge and Action.* New York: G.P. Putnam's Sons.

Dickson, David (1988) *The New Politics of Science.* Chicago: University of Chicago Press.

Dobbs, Betty Jo Teeter and Margaret Jacob (1995) *Newton and the Culture of Newtonianism.* Atlantic Highlands, NJ: Humanities Press.

Doing, Park (2007) "Give Me a Laboratory and I Will Raise a Discipline: The Past, Present, and Future Politics of Laboratory Studies in STS." In E. J. Hackett, O. Amsterdamska, M. Lynch and J. Wajcman, eds, *The Handbook of Science and Technology Studies,* 3rd edn. Cambridge, MA: MIT Press, 279–95.

Dornan, Christopher (1990) "Some Problems in Conceptualizing the Issue of Science and the Media." *Critical Studies in Mass Communication* 7: 48–71.

Downer, John (2007) "When the Chick Hits the Fan: Representativeness and Reproducibility in Technological Tests." *Social Studies of Science* 37: 7–26.

Dreyfus, Hubert L. (1972) *What Computers Can't Do: A Critique of Artificial Reason.* New York: Harper & Row.

Dreyfuss, Rochelle Cooper (1995) "Is Science a Special Case? The Admissability of Scientific Evidence After *Daubert v. Merrell Dow.*" *Texas Law Review* 73: 1779–804.

Dumit, Joseph (2003) *Picturing Personhood: Brain Scans and Biomedical Identity.* Princeton: Princeton University Press.

Duncker, Elke (2001) "Symbolic Communication in Multidisciplinary Cooperations." *Science, Technology, & Human Values* 26: 349–86.

Dupré, John (1993) *The Disorder of Things: Metaphysical Foundations of the Disunity of Science.* Cambridge, MA: Harvard University Press.

Durant, Darrin (2008) "Accounting for Expertise: Wynne and the Autonomy of the Lay Public Actor." *Public Understanding of Science* 17: 5–20.

Easlea, Brian (1986) "The Masculine Image of Science with Special Reference to Physics: How Much does Gender Really Matter?" in J. Harding, ed., *Perspectives on Gender and Science.* London: The Falmer Press, 132–59.

Edge, David (1979) "Quantitative Measures of Communication in Science: A Critical Review." *History of Science* 17: 102–34.

Edwards, Paul N. (1997) *The Closed World: Computers and the Politics of Discourse in Cold War America.* Cambridge: MIT Press.

Edwards, Paul N. (1999) "Data-Laden Models, Model-Filtered Data: Uncertainty and Politics in Global Climate Science." *Science as Culture* 8: 437–72.

Einsiedel, Edna, Erling Jelsøe, and Thomas Breck (2001) "Publics at the Technology Table: The Consensus Conference in Denmark, Canada, and Australia." *Public Understanding of Science* 10: 83–93.

Ellul, Jacques (1964) *The Technological Society.* Tr. J. Wilkinson. New York: Knopf.

Epstein, Steven (1996) *Impure Science: AIDS, Activism, and the Politics of Knowledge.* Berkeley: University of California Press.

Epstein, Steven (2007) *Inclusion: The Politics of Difference in Medical Research.* Chicago: University of Chicago Press.

Escobar, Arturo (1999) "After Nature." *Current Anthropology* 40: 1–16.

Etzkowitz, Henry, Carol Kemelgor, and Brian Uzzi (2000) *Athena Unbound: The Advancement of Women in Science and Technology.* Cambridge: Cambridge University Press.

Etzkowitz, Henry and Loet Leydesdorff, eds (1997) *Universities and the Global Economy: A Triple Helix of University–Industry–Government Relations.* London: Pinter.

European Commission Directorate-General for Research (2006) *She Figures 2006: Women in Science, Statistics and Indicators.* Available at: http://ec.europa.eu/research/science-society/pdf/she_figures_2006_en.pdf.

Evans, Robert (1997) "Soothsaying or Science? Falsification, Uncertainty and Social Change in Macroeconomic Modelling." *Social Studies of Science* 27: 395–438.

Ezrahi, Yaron (1990) *The Descent of Icarus: Science and the Transformation of Contemporary Democracy.* Cambridge, MA: Harvard University Press.

Farkas, Nicole (1999) "Dutch Science Shops: Matching Community Needs with University R&D." *Science Studies* 12: 33–47.

Farley, John and Gerald Geison (1974) "Science, Politics and Spontaneous Generation in Nineteenth-Century France: The Pasteur–Pouchet Debate." *Bulletin of the History of Medicine* 48: 161–98.

Faulkner, Wendy (2000) "Dualisms, Hierarchies and Gender in Engineering." *Social Studies of Science* 30: 759–92.

Faulkner, Wendy (2007) " 'Nuts and Bolts and People': Gender-Troubled Engineering Identities." *Social Studies of Science* 37: 331–56.

Fausto-Sterling, Anne (1985) *Myths of Gender: Biological Theories about Women and Men.* New York: Basic Books.

Findlen, Paula (1993) "Controlling the Experiment: Rhetoric, Court Patronage and the Experimental Method of Francisco Redi." *History of Science* 31: 35–64.

Fine, Arthur (1986) *The Shaky Game: Einstein, Realism and the Quantum Theory.* Chicago: University of Chicago Press.

Fischer, Corinna, Loet Leydesdorff, and Malte Schophaus (2004) "Science Shops in Europe: The Public as Stakeholder." *Science & Public Policy* 31: 199–211.

Fisher, Jill (2009) *Medical Research for Hire: The Political Economy of Pharmaceutical Clinical Trials.* New Brunswick, NJ: Rutgers University Press.

Fishman, Jennifer (2004) "Manufacturing Desire: The Commodification of Female Sexual Dysfunction." *Social Studies of Science* 34(2): 187–218.

Fitzgerald, Deborah (1993) "Farmers Deskilled; Hybrid Corn and Farmers' Work." *Technology and Culture* 34: 324–43.

Fleck, Ludwik (1979) *Genesis and Development of a Scientific Fact* (first published 1935). Ed. T. J. Trenn and R. K. Merton. Tr. F. Bradley and T. J. Trenn. Chicago: University of Chicago Press.

Fodor, Jerry (1988) "A Reply to Churchland's 'Perceptual Plasticity and Theoretical Neutrality.' " *Philosophy of Science* 55: 188–98.

Forsythe, Diana E. (2001) *Studying Those Who Study Us: An Anthropologist in the World of Artificial Intelligence.* Tr. David J. Hess. Stanford, CA: Stanford University Press.

Fox, Mary Frank (1983) "Publication Productivity among Scientists: A Critical Review." *Social Studies of Science* 13: 285–305.

Fox, Mary Frank (1991) "Gender, Environmental Milieu, and Productivity in Science." In H. Zuckerman, J. R. Cole, and J. T. Bruer, eds, *The Outer Circle: Women in the Scientific Community.* New York: W. W. Norton, 188–204.

Franklin, Allan (1997) "Calibration." *Perspectives on Science* 5: 31–80.

Franklin, Ursula (1990) *The Real World of Technology.* Concord, ON: Anansi.

Fraser, Nancy and Linda J. Nicholson (1990) "Social Criticism without Philosophy: An Encounter between Feminism and Postmodernism." In L. J. Nicholson, ed., *Feminism/Postmodernism.* New York: Routledge, 19–38.

Friedman, Michael (1999) *Reconsidering Logical Positivism.* Cambridge: Cambridge University Press.

Fuchs, Stephan (1992) *The Professional Quest for Truth: A Social Theory of Science and Knowledge.* Albany, NY: SUNY Press.

Fuchs, Stephan (1993) "Positivism is the Organizational Myth of Science." *Perspectives on Science* 1: 1–23.

Fujimura, Joan H. (1988) "The Molecular Biological Bandwagon in Cancer Research: Where Social Worlds Meet." *Social Problems* 35: 261–83.

Fuller, Steve (2000) *Thomas Kuhn: A Philosophical History for Our Times.* Chicago: University of Chicago Press.

Fuller, Steve (2007) *The Knowledge Book: Key Concepts in Philosophy, Science and Culture.* Montreal: McGill-Queen's University Press.

Futrell, Robert (2003) "Technical Adversarialism and Participatory Collaboration in the U.S. Chemical Weapons Disposal Program." *Science, Technology & Human Values* 28: 451–82.

Gale, George and Cassandra L. Pinnick (1997) "Stalking Theoretical Physicists: An Ethnography Flounders: A Response to Merz and Knorr Cetina." *Social Studies of Science* 27: 113–23.

Galison, Peter (1997) *Image and Logic: A Material Culture of Microphysics.* Chicago: University of Chicago Press.

Galison, Peter and David J. Stump (1996) "The Disunity of Science: Boundaries, Contexts, and Power." Stanford: Stanford University Press.

Garber, Daniel (1995) "Experiment, Community, and the Constitution of Nature in the Seventeenth Century." *Perspectives on Science* 3: 173–205.

Gergen, Kenneth (1986) "Correspondence versus Autonomy in the Language of Understanding Human Action." In D. Fiske and R. Shweder, eds, *Metatheory in Social Science: Pluralisms and Subjectivities.* Chicago: University of Chicago Press, 132–62.

Ghamari-Tabrizi, Sharon (2005) *The Worlds of Herman Kahn: The Intuitive Science of Thermonuclear War.* Cambridge, MA: Harvard University Press.

Gibbons, Michael, Camille Limoges, Helga Nowotny, Simon Schwartzmann, Peter Scott, and Michael Trow (1994) *The New Production of Knowledge: The Dynamics of Science and Research in Contemporary Societies*. London: Sage.

Gieryn, Thomas F. (1992) "The Ballad of Pons and Fleischmann: Experiment and Narrativity in the (Un)Making of Cold Fusion." In E. McMullin, ed., *The Social Dimensions of Science*. Notre Dame, IN: University of Notre Dame Press, 217–43.

Gieryn, Thomas (1996) "Policing STS: A Boundary-Work Souvenir from the Smithsonian Exhibition on 'Science in American Life'." *Science, Technology & Human Values* 21: 100–15.

Gieryn, Thomas (1999) *Cultural Boundaries of Science: Credibility on the Line*. Chicago: University of Chicago Press.

Gieryn, Thomas (2002) "What Buildings Do." *Theory and Society* 31: 35–74.

Gieryn, Thomas F. and Anne E. Figert (1986) "Scientists Protect their Cognitive Authority: The Status Degradation Ceremony of Sir Cyril Burt." In G. Böhme and N. Stehr, eds, *The Knowledge Society: The Growing Impact of Scientific Knowledge on Social Relations*. Dordrecht: D. Reidel, 67–86.

Gieryn, Thomas F. and Anne E. Figert (1990) "Ingredients for a Theory of Science in Society: O-Rings, Ice Water, C-Clamp, Richard Feynman, and The Press." In S. E. Cozzens and T. F. Gieryn, eds, *Theories of Science in Society*. Bloomington: Indiana University Press, 67–97.

Gilbert, G. Nigel (1977) "Referencing as Persuasion." *Social Studies of Science* 7: 113–22.

Gilbert, G. Nigel and Michael Mulkay (1984) *Opening Pandora's Box: A Sociological Analysis of Scientists' Discourse*. Cambridge: Cambridge University Press.

Gilligan, Carol (1982) *In a Different Voice: Psychological Theory and Women's Development*. Cambridge, MA: Harvard University Press.

Gillespie, Tarleton (2007) *Wired Shut: Copyright and the Shape of Digital Culture*. Cambridge, MA: MIT Press.

Gillispie, Charles Coulton (1960) *The Edge of Objectivity: An Essay in the History of Scientific Ideas*. Princeton, NJ: Princeton University Press.

Gingras, Yves and Michel Trepanier (1993) "Constructing a Tokamak: Political, Economic and Technical Factors as Constraints and Resources." *Social Studies of Science* 23: 5–36.

Ginsburg, Faye E. and Rayna Rapp, eds (1995) *Conceiving the New World Order: The Global Politics of Reproduction*. Berkeley: University of California Press.

Glover, Judith (2000) *Women and Scientific Employment*. New York: St. Martin's Press.

Godin, Benoit (2006) "The Linear Model of Innovation: The Historical Construction of an Analytical Framework." *Science, Technology & Human Values* 31: 639–67.

Goodfield, June (1981) *An Imagined World: A Story of Scientific Discovery*. New York: Harper & Row.

Goodman, Nelson (1983) *Fact, Fiction, and Forecast*, 4th edn (first published 1954). Cambridge, MA: Harvard University Press.

Goodwin, Charles (1995) "Seeing in Depth." *Social Studies of Science* 25: 237–74.

Goonatilake, Susantha (1993) "Modern Science and the Periphery." In S. Harding, ed., *The "Racial" Economy of Science*. Bloomington: Indiana University Press, 259–67.

Goven, Joanna (2003) "Deploying the Consensus Conference in New Zealand: Democracy and De-problematization." *Public Understanding of Science* 12: 423–40.

Gregory, Jane and Steve Miller (1998) *Science in Public: Communication, Culture, and Credibility*. New York: Plenum.

Grint, Keith and Steve Woolgar (1997) *The Machine at Work: Technology, Work and Organization*. Cambridge: Polity Press.

Gross, Alan G. (1990a) *The Rhetoric of Science*. Cambridge, MA: Harvard University Press.

Gross, Alan G. (1990b) "The Origin of Species: Evolutionary Taxonomy as an Example of the Rhetoric of Science." In H. W. Simons, ed., *The Rhetorical Turn: Invention and Persuasion in the Conduct of Inquiry*. Chicago: Chicago University Press, 91–115.

Grundmann, Reiner and Nico Stehr (2000) "Social Science and the Absence of Nature: Uncertainty and the Reality of Extremes." *Social Science Information* 39: 155–79.

Guice, Jon (1998) "Controversy and the State: Lord ARPA and Intelligent Computing." *Social Studies of Science* 28: 103–38.

Guillory, John (2002) "The Sokal Affair and the History of Criticism." *Critical Inquiry* 28: 470–508.

Gunnarsdóttir, Kristrún (2005) "Scientific Journal Publications: On the Role of Electronic Preprint Exchange in the Distribution of Scientific Literature." *Social Studies of Science* 35: 549–79.

Gupta, Namrata (2007) "Indian Women in Doctoral Education in Science and Engineering: A Study of Informal Milieu at the Reputed Indian Institutes of Technology." *Science, Technology & Human Values* 32: 507–33.

Gusterson, Hugh (1996) *Nuclear Rites: A Weapons Laboratory at the End of the Cold War*. Berkeley: University of California Press.

Guston, David (1999a) "Changing Explanatory Frameworks in the U.S. Government's Attempt to Define Research Misconduct." *Science and Engineering Ethics* 5: 137–54.

Guston, David (1999b) "Stabilizing the Boundary between US Politics and Science: The Rôle of the Office of Technology Transfer as a Boundary Organization." *Social Studies of Science* 29: 87–112.

Hacker, Sally (1990) *Doing it the Hard Way: Investigations of Gender and Technology*. Boston: Unwin Hyman.

Hacking, Ian (1983) *Representing and Intervening: Introductory Topics in the Philosophy of Natural Science*. Cambridge: Cambridge University Press.

Hacking, Ian (1999) *The Social Construction of What?* Cambridge, MA: Harvard University Press.

Hamlett, Patrick W. (2003) "Technology Theory and Deliberative Democracy." *Science, Technology & Human Values* 28(1): 112–40.

Hanson, Norwood Russell (1958) *Patterns of Discovery: An Inquiry into the Conceptual Foundations of Science*. Cambridge: Cambridge University Press.

Haraway, Donna (1976) *Crystals, Fabrics, and Fields: Metaphors of Organicism in 20th Century Developmental Biology*. New Haven: Yale University Press.

Haraway, Donna (1985) "A Manifesto for Cyborgs: Science, Technology, and Socialist Feminism in the 1980s." *Socialist Review* 80: 65–107.

Haraway, Donna (1988) "Situated Knowledges: The Science Question in Feminism and the Privilege of Partial Perspective." *Feminist Studies* 14: 575–609.

Haraway, Donna (1989) *Primate Visions: Gender, Race, and Nature in the World of Modern Science*. New York: Routledge.

Hård, Mikael (1993) "Beyond Harmony and Consensus: A Social Conflict Approach to Technology." *Science, Technology & Human Values* 18: 408–432.

Harding, Sandra (1986) *The Science Question in Feminism*. Ithaca, NY: Cornell University Press.

Harding, Sandra (1991) *Whose Science? Whose Knowledge?: Thinking from Women's Lives*. Ithaca: Cornell University Press.

Harding, Sandra (2006) *Science and Social Inequality: Feminist and Postcolonial Issues*. Urbana: University of Illinois Press.

Harkness, Deborah E. (1997) "Managing an Experimental Household: The Dees of Mortlake and the Practice of Natural Philosophy." *Isis* 88: 247–62.

Hartsock, Nancy C. M. (1983) "The Feminist Standpoint: Developing a Ground for a Specifically Feminist Historical Materialism." In S. Harding and M. Hintikka, eds, *Feminist Perspectives on Epistemology, Metaphysics, Methodology and Philosophy of Science*, Dordrecht: D. Reidel, 283–310.

Harwood, Jonathan (1976, 1977) "The Race–Intelligence Controversy: A Sociological Approach." *Social Studies of Science* 6–7: 369–94, 1–30.

Hayden, Cori (2003) *When Nature Goes Public: The Making and Unmaking of Bioprospecting in Mexico*. Princeton: Princeton University Press.

Hayles, N. Katherine (1999) *How We Became Posthuman: Virtual Bodies in Cybernetics, Literature, and Informatics*. Chicago: University of Chicago Press.

Headrick, Daniel R. (1988) *The Tentacles of Progress: Technology Transfer in the Age of Imperialism, 1850–1940*. Oxford: Oxford University Press.

Healy, David and Dinah Cattell (2003) "Interface between Authorship, Industry and Science in the Domain of Therapeutics." *British Journal of Psychiatry* 183: 22–7.

Hecht, Gabrielle (2002) "Rupture-Talk in the Nuclear Age: Conjugating Colonial Power in Africa." *Social Studies of Science* 32: 691–727.

Heidegger, Martin (1977) *The Question Concerning Technology, and Other Essays*. Tr. W. Lovitt (first published 1954). New York: Harper & Row.

Heilbroner, Robert L. (1994) "Do Machines Make History?" In M. R. Smith and L. Marx, eds, *Does Technology Drive History? The Dilemma of Technological Determinism*. Cambridge, MA: MIT Press, 53–65.

Henderson, Kathryn (1991) "Flexible Sketches and Inflexible Data Bases: Visual Communication, Conscription Devices, and Boundary Objects in Design Engineering." *Science, Technology, & Human Values* 16: 448–73.

Henke, Christopher R. (2000) "Making a Place for Science: The Field Trial." *Social Studies of Science* 30: 483–511.

Hermanowicz, Joseph C. (2006) "What Does It Take to Be Successful?" *Science, Technology, & Human Values* 31: 135–52.

Hess, David J. (1997) *Science Studies: An Advanced Introduction*. New York: New York University Press.

Hesse, Mary B. (1966) *Models and Analogies in Science*. Notre Dame, IN: University of Notre Dame Press.

Higgitt, Rebekah and Charles W. J. Withers (2008) "Science and Sociability: Women as Audience at the British Association for the Advancement of Science, 1831–1901." *Isis* 90: 1–27.

Hilgartner, Stephen (1990) "The Dominant View of Popularization: Conceptual Problems, Political Uses." *Social Studies of Science* 20: 519–39.

Hoffman, Robert R. (1985) "Some Implications of Metaphor for Philosophy and Psychology of Science." In W. Paprotté and R. Rirven, eds, *The Ubiquity of Metaphor: Metaphor in Language and Thought*. Amsterdam: John Benjamin, 327–80.

Hoffmann, Roald and Shira Leibowitz (1991) "Molecular Mimicry, Rachel and Leah, the Israeli Male, and the Inescapable Metaphor of Science." *Michigan Quarterly Review* 30: 383–98.

Hong, Sungook (1998) "Unfaithful Offspring? Technologies and Their Trajectories." *Perspectives on Science* 6: 259–87.

Hubbard, Ruth, Mary Sue Henifin, and Barbara Fried (1979) *Women Look at Biology Looking at Women: A Collection of Feminist Critiques*. Cambridge, MA: Schenkman.

Huber, Peter (1994) *Galileo's Revenge: Junk Science in the Courtroom*. New York: Basic Books.

Hughes, T. P. (1985) "Edison and Electric Light." In D. Mackenzie and J. Wajcman, eds, *The Social Shaping of Technology: How the Refrigerator got its Hum*. Milton Keynes: Open University Press, 39–52.

Hughes, Thomas P. (1987) "The Evolution of Large Technological Systems." In W. E. Bijker, T. P. Hughes, and T. Pinch, eds, *The Social Construction of Technological Systems*. Cambridge, MA: MIT Press, 51–82.

Hull, David L. (1988) *Science as a Process: An Evolutionary Account of the Social and Conceptual Development of Science*. Chicago: University of Chicago Press.

Irwin, Alan (1995) *Citizen Science: A Study of People, Expertise and Sustainable Development*. London: Routledge.

Irwin, Alan (2008) "STS Perspectives on Scientific Governance." In E. J. Hackett, O. Amsterdamska, M. Lynch and J. Wajcman, eds, *The Handbook of Science and Technology Studies*, 3rd edn. Cambridge, MA: MIT Press, 583–608.

Jacob, Margaret C. (1976) *The Newtonians and the English Revolution, 1689–1720.* Ithaca, NY: Cornell University Press.

Jasanoff, Sheila (1995) *Science at the Bar: Law, Science, and Technology in America.* Cambridge, MA: Harvard University Press.

Jasanoff, Sheila (1996) "Beyond Epistemology: Relativism and Engagement in the Politics of Science." *Social Studies of Science* 26: 393–418.

Jasanoff, Sheila (1987) "Contested Boundaries in Policy-Relevant Science." *Social Studies of Science* 17: 195–230.

Jasanoff, Sheila (2005) *Designs on Nature: Science and Democracy in Europe and the United States.* Princeton: Princeton University Press.

Jasper, James M. (1992) "Three Nuclear Energy Controversies." In D. Nelkin, ed., *Controversy*, Vol. 3. London: Sage, 97–111.

Jones, Roger S. (1982) *Physics as Metaphor.* Minneapolis: University of Minnesota Press.

Jordan, Kathleen and Michael Lynch (1992) "The Sociology of a Genetic Engineering Technique: Ritual and Rationality in the Performance of the 'Plasmid Prep'." In A. E. Clark and J. H. Fujimura, eds, *The Right Tools for the Job: At Work in Twentieth Century Life.* Princeton: Princeton University Press, 77–114.

Joyce, Kelly A. (2008) *Magnetic Appeal: MRI and the Myth of Transparency.* Ithaca, NY: Cornell University Press.

Kaiser, David (1994) "Bringing the Human Actors Back on Stage: The Personal Context of the Einstein–Bohr Debate." *British Journal for the History of Science* 27: 127–52.

Kaiser, David (2005) *Drawing Theories Apart: The Dispersion of Feynman Diagrams in Postwar Physics.* Chicago: University of Chicago Press.

Kaiserfeld, Thomas (1996) "Computerizing the Swedish Welfare State: The Middle Way of Technological Success and Failure." *Technology and Culture* 37: 249–79.

Kay, Lily E. (1995) "Who Wrote the Book of Life? Information and the Transformation of Molecular Biology, 1945–55." *Science in Context* 8: 609–34.

Keller, Evelyn Fox (1983) *A Feeling for the Organism: The Life and Work of Barbara McClintock.* New York: W. H. Freeman.

Keller, Evelyn Fox (1985) *Reflections on Gender and Science.* New Haven, CT: Yale University Press.

Kevles, Daniel J. (1998) *The Baltimore Case: A Trial of Politics, Science, and Character.* New York: W. W. Norton.

Kidder, Tracy (1981) *The Soul of a New Machine.* New York: Avon Books.

Kiernan, Vincent (1997) "Ingelfinger, Embargoes, and Other Controls on the Dissemination of Science News." *Science Communication* 18: 297–319.

Kim, Jongyoung (2009) "Public Feeling for Science: The Hwang Affair and Hwang Supporters." *Public Understanding of Science*, forthcoming.

Kim, Leo (2008) "Explaining the Hwang Scandal: National Scientific Culture and its Global Relevance." *Science as Culture* 17: 397–415.

Kirsch, David A. (2000) *The Electric Vehicle and the Burden of History*. New Brunswick, NJ: Rutgers University Press.

Kitcher, Phillip (1991) "Persuasion." In M. Pera and W. R. Shea, eds, *Persuading Science: The Art of Scientific Rhetoric*. Canton, MA: Science History Publications, 3–27.

Kitcher, Philip (2001) *Science, Truth, and Democracy*. Oxford: Oxford University Press.

Kitzinger, Jenny (2008) "Questioning Hype, Rescuing Hope? The Hwang Stem Cell Scandal and the Reassertion of Hopeful Horizons." *Science as Culture* 17: 417–34.

Kleif, Tine and Wendy Faulkner (2003) " 'I'm No Athlete [but] I Can Make This Thing Dance!' – Men's Pleasures in Technology." *Science, Technology, & Human Values* 28: 296–325.

Kleinman, Daniel Lee (1998) "Untangling Context: Understanding a University Laboratory in the Commercial World." *Science, Technology & Human Values* 23: 285–314.

Kleinman, Daniel Lee and Abby Kinchy (2003) "Organizing Credibility: Structural Considerations on the Borders of Ecology and Politics." *Social Studies of Science* 33: 869–96.

Kline, Ronald (1992) *Steinmetz: Engineer and Socialist*. Baltimore: Johns Hopkins University Press.

Kline, Ronald and Trevor Pinch (1996) "Users as Agents of Technological Change: The Social Construction of the Automobile in the Rural United States." *Technology and Culture* 37: 763–95.

Kling, Rob (1992) "Audiences, Narratives, and Human Values in Social Studies of Technology." *Science, Technology & Human Values* 17: 349–65.

Klintman, Mikael (2002) "The Genetically Modified (GM) Food Labelling Controversy: Ideological and Epistemic Crossovers." *Social Studies of Science* 32: 71–91.

Knorr Cetina, Karin D. (1977) "Producing and Reproducing Knowledge: Descriptive or Constructive?" *Social Science Information* 16: 669–96.

Knorr Cetina, Karin D. (1979) "Tinkering Toward Success: Prelude to a Theory of Scientific Practice." *Theory and Society* 8: 347–76.

Knorr Cetina, Karin D. (1981) *The Manufacture of Knowledge: An Essay on the Constructivist and Contextual Nature of Science*. Oxford: Pergamon Press.

Knorr Cetina, Karin D. (1983) "The Ethnographic Study of Scientific Work: Towards a Constructivist Interpretation of Science." In K. D. Knorr Cetina and M. Mulkay, eds, *Science Observed: Perspectives on the Social Study of Science*. London: Sage, 115–40.

Knorr Cetina, Karin D. (1999) *Epistemic Cultures: How the Sciences Make Knowledge*. Cambridge, MA: Harvard University Press.

Knorr Cetina, Karin D. and Aaron V. Cicourel (1981) *Advances in Social Theory and Methodology: Toward an Integration of Micro- and Macro-Sociologies*. Boston, MA: Routledge & Kegan Paul.

Kohler, Robert E. (1994) *Lords of the Fly: Drosophila Genetics and the Experimental Life*. Chicago: University of Chicago Press.

Koshland, Daniel E. (1994) "Scientific Evidence in Court." *Science* 266: 1787.

Kranakis, Eda (2007) "Patents and Power: European Patent-System Integration in the Context of Globalization." *Technology and Culture* 48: 689–728.

Kripke, Saul A. (1982) *Wittgenstein on Rules and Private Language.* Cambridge, MA: Harvard University Press.

Kuhn, Thomas S. (1970) *The Structure of Scientific Revolutions,* 2nd edn (1st edn 1962). Chicago: Chicago University Press.

Kuhn, Thomas S. (1970a) "Reflections on my Critics." In I. Lakatos and A. Musgrave, eds, *Criticism and the Growth of Knowledge.* Cambridge: Cambridge University Press, 231–78.

Kuhn, Thomas S. (1977) "Second Thoughts on Paradigms." In F. Suppe, ed., *The Structure of Scientific Theories,* 2nd edn. Urbana: University of Illinois Press, 459–82.

Kuklick, Henrika (1991) "Contested Monuments: The Politics of Archaeology in Southern Africa." In G. Stocking Jr., ed., *Colonial Situations: Essays on the Contextualization of Ethnographic Knowledge.* Madison: University of Wisconsin Press, 135–69.

Kula, Witold (1986) *Measures and Men.* Tr. R. Szreter. Princeton: Princeton University Press.

Kusch, Martin (2004) "Rule-Scepticism and the Sociology of Scientific Knowledge: The Bloor–Lynch Debate Revisited." *Social Studies of Science* 34: 571–91.

Lagesen, Vivian Anette (2007) "The Strength of Numbers: Strategies to Include Women into Computer Science." *Social Studies of Science* 37: 67–92.

Laird, Frank (1993) "Participatory Analysis, Democracy, and Technological Decision Making." *Science, Technology & Human Values* 18: 341–61.

Lakatos, Imre (1976) *Proofs and refutations: The Logic of Mathematical Discovery.* Ed. J. Worrall and E. Zahar. Cambridge: Cambridge University Press.

Langford, Jean M. (2002) *Fluent Bodies: Ayurvedic Remedies for Postcolonial Imbalance.* Durham: Duke University Press.

Langwick, Stacey A. (2007) "Devils, Parasites, and Fierce Needles: Healing and the Politics of Translation in Southern Tanzania." *Science, Technology & Human Values* 32: 88–117.

Latour, Bruno (1983) "Give Me a Laboratory and I Will Raise the World." In K. D. Knorr Cetina and M. Mulkay, eds, *Science Observed: Perspectives on the Social Study of Science.* London: Sage, 141–70.

Latour, Bruno (1987) *Science in Action: How to Follow Scientists and Engineers through Society.* Cambridge, MA: Harvard University Press.

Latour, Bruno (1988) *The Pasteurization of France.* Tr. A. Sheridon and J. Law. Cambridge, MA: Harvard University Press.

Latour, Bruno (1990) "The Force and the Reason of Experiment." In H. E. L. Grand, *Experimental Inquiries.* Dordrecht: Kluwer, 49–80.

Latour, Bruno (1993) *We Have Never Been Modern.* Tr. C. Porter. New York: Harvester Wheatsheaf.

Latour, Bruno (1994) "Pragmatogonies: A Mythical Account of How Humans and Nonhumans Swap Properties." *American Behavioral Scientist* 37: 791–808.

Latour, Bruno (1999) *Pandora's Hope: Essays on the Reality of Science Studies.* Cambridge, MA: Harvard University Press.

Latour, Bruno (2004) *Politics of Nature: How to Bring the Sciences into Democracy* Cambridge, MA: Harvard University Press.

Latour, Bruno (2005) *Reassembling the Social: An Introduction to Actor-Network-Theory.* Oxford: Oxford University Press.

Latour, Bruno and Steve Woolgar (1979) *Laboratory Life: The Social Construction of Scientific Facts.* Beverly Hills: Sage.

Latour, Bruno and Steve Woolgar (1986) *Laboratory Life: The Construction of Scientific Facts,* 2nd edn. Princeton: Princeton University Press.

Laudan, Rachel (1984) "Cognitive Change in Technology and Science." In R. Laudan, ed., *The Nature of Technological Knowledge: Are Models of Scientific Change Relevant.* Dordrecht: D. Reidel, 83–104.

Law, John (1987) "Technology and Heterogeneous Engineering: The Case of Portuguese Expansion." In W. E. Bijker, T. P. Hughes, and T. J. Pinch, eds, *The Social Construction of Technological Systems: New Directions in the Sociology and History of Technology.* Cambridge, MA: MIT Press, 111–34.

Law, John (1999) "After ANT: Complexity, Naming and Topology." In J. Law and J. Hassard, eds, *Actor Network Theory and After.* Oxford: Blackwell, 1–14.

Layton, Edwin (1971) "Mirror-Image Twins: The Communities of Science and Technology in 19th-Century America." *Technology and Culture* 12: 562–80.

Layton, Edwin (1974) "Technology as Knowledge." *Technology and Culture* 15: 31–41.

Le Grand, H. E. (1986) "Steady as a Rock: Methodology and Moving Continents." In J. A. Schuster and R. R. Yeo, eds, *The Politics and Rhetoric of Scientific Method.* Dordrecht: D. Reidel, 97–138.

Leem, So Yeon and Jin Hee Park (2008) "Rethinking Women and Their Bodies in the Age of Biotechnology: Feminist Commentaries on the Hwang Affair." *East Asian Science, Technology, and Society* 2: 9–26.

Leiss, William (1972) *The Domination of Nature.* New York: George Braziller.

Lengwiler, Martin (2008) "Participatory Approaches in Science and Technology: Historical Origins and Current Practices in Critical Perspective." *Science, Technology & Human Values* 33(2): 186–200.

Leplin, Jarrett, ed. (1984) *Scientific Realism.* Berkeley: University of California Press.

Lewenstein, Bruce (1995) "From Fax to Facts: Communication in the Cold Fusion Saga." *Social Studies of Science* 25: 403–36.

Lightfield, E. Timothy (1971) "Output and Recognition of Sociologists." *The American Sociologist* 6: 128–33.

Livingstone, David N. (2003) *Putting Science in Its Place: Geographies of Scientific Knowledge.* Chicago: University of Chicago Press.

Locke, David (1992) *Science as Writing,* 1st edn. New Haven, CT: Yale University Press.

Locke, David (2002) "The Public Understanding of Science – A Rhetorical Invention." *Science, Technology & Human Values* 27: 87–111.

Long, J. Scott (2001) *From Scarcity to Visibility: Gender Differences in the Careers of Doctoral Scientists and Engineers.* Washington, DC: National Academy Press.

Long, J. Scott and R. McGinnis (1981) "Organizational Context and Scientific Productivity." *American Sociological Review* 46: 422–42.

Longino, Helen E. (1990) *Science as Social Knowledge: Values and Objectivity in Scientific Inquiry.* Princeton: Princeton University Press.

Longino, Helen E. (2002) *The Fate of Knowledge.* Princeton: Princeton University Press.

Lowncy, Kathleen S. (1998) "Floral Entrepreneurs: Kudzu as Agricultural Solution and Ecological Problem." *Sociological Spectrum* 18: 93–114.

Lynch, Michael (1985) *Art an Artifact in Laboratory Science: A Study of Shop Work and Shop Talk in a Research Laboratory.* London: Routledge and Kegan Paul.

Lynch, Michael (1990) "The Externalized Retina: Selection and Mathematization in the Visual Documentation of Objects in the Life Sciences." In M. Lynch and S. Woolgar, eds, *Representation in Scientific Practice.* Cambridge, MA: The MIT Press, 153–86.

Lynch, Michael (1991) "Laboratory Space and the Technological Complex: An Investigation of Topical Contextures." *Science in Context* 4: 51–78.

Lynch, Michael (1992a) "Extending Wittgenstein: The Pivotal Move from Epistemology to the Sociology of Science." In A. Pickering, ed., *Science as Practice and Culture.* Chicago: University of Chicago Press, 215–65.

Lynch, Michael (1992b) "From the 'Will to Theory' to the Discursive Collage: A Reply to Bloor's 'Left and Right Wittgensteinians'." In A. Pickering, ed., *S cience as Practice and Culture.* Chicago: University of Chicago Press, 283–300.

Lynch, Michael (1993) *Scientific Practice and Ordinary Action: Ethnomethodology and Social Studies of Science.* Cambridge: Cambridge University Press.

Lynch, Michael (1998) "The Discursive Production of Uncertainty: The OJ Simpson 'Dream Team' and the Sociology of Knowledge Machine." *Social Studies of Science* 28: 829–68.

Lynch, William and Ronald Kline (2000) "Engineering Practice and Engineering Ethics." *Science, Technology & Human Values* 25: 195–225.

MacArthur, Robert H. and Edward O. Wilson (1967) *The Theory of Island Biogeography.* Princeton: Princeton University Press.

MacKenzie, Donald (1978) "Statistical Theory and Social Interests: A Case Study." *Social Studies of Science* 8: 35–83.

MacKenzie, Donald (1981) *Statistics in Britain, 1865–1930: The Social Construction of Scientific Knowledge.* Edinburgh: Edinburgh University Press.

MacKenzie, Donald (1989) "From Kwajalein to Armageddon? Testing and the Social Construction of Missile Accuracy." In D. Gooding, T. Pinch, and S. Schaffer, eds, *The Uses of Experiment: Studies in the Natural Sciences.* Cambridge: Cambridge University Press, 409–35.

MacKenzie, Donald (1990) *Inventing Accuracy: A Historical Sociology of Nuclear Missile Guidance.* Cambridge, MA: MIT Press.

MacKenzie, Donald (1999) "Slaying the Kraken: The Sociohistory of a Mathematical Proof." *Social Studies of Science* 29: 7–60.

MacKenzie, Donald (2006) *An Engine, Not a Camera: How Financial Models Shape Markets.* Cambridge, MA: MIT Press.

MacRoberts, M. H. and Barbara R. MacRoberts (1996) "Problems of Citation Analysis." *Scientometrics* 36: 435–44.

Maines, Rachel (2001) "Socially Camouflaged Technologies: The Case of the Electromechanical Vibrator." In M. Wyer, M. Barbercheck, D. Giesman, H. Öztürk, and M. Wayne, eds, *Women, Science, and Technology: A Reader in Feminist Science Studies.* New York: Routledge, 271–85.

Marie, Jenny (2008) "For Science, Love and Money: The Social Worlds of Poultry and Rabbit Breeding in Britain, 1900–1940." *Social Studies of Science* 38: 919–36.

Martin, Brian (1991) *Scientific Knowledge in Controversy: The Social Dynamics of the Fluoridation Debate.* Albany, NY: SUNY Press.

Martin, Brian (1996) "Sticking a Needle into Science: The Case of Polio Vaccines and the Origin of AIDS." *Social Studies of Science* 26: 245–76.

Martin, Brian (2006) "Strategies for Alternative Science." In S. Frickel and K. Moore, eds, *The New Political Sociology of Science: Institutions, Networks, and Power.* Madison: University of Wisconsin Press, 272–98.

Martin, Emily (1991) "The Egg and the Sperm: How Science has Constructed a Romance Based on Stereotypical Male–Female Roles." *Signs* 16: 485–501.

Masco, Joseph (2006) *The Nuclear Borderlands: The Manhattan Project in Post-Cold War New Mexico.* Princeton: Princeton University Press.

Maynard, Douglas W. and Nora Cate Shaeffer (2000) "Toward a Sociology of Social Scientific Knowledge: Survey Research and Ethnomethodology's Asymmetric Alternates." *Social Studies of Science* 30: 323–70.

McSherry, Corynne (2001) *Who Owns Academic Work? Battling for Control of Intellectual Property.* Cambridge, MA: Harvard University Press.

Medawar, Peter (1963) "Is the Scientific Paper a Fraud?" *The Listener* 70 (1798): 377–78.

Mellor, Felicity (2003) "Between Fact and Fiction: Demarcating Science from Non-Science in Popular Physics Books." *Social Studies of Science* 33: 509–38.

Mellor, Felicity (2007) "Colliding Worlds: Asteroid Research and the Legitimization of War in Space." *Social Studies of Science* 37: 499–531.

Mendelsohn, Everett (1977) "The Social Construction of Scientific Knowledge." In E. Mendelsohn, P. Weingart, and R. Whitley, eds, *The Social Production of Scientific Knowledge.* Dordrecht: D. Reidel, 3–26.

Merchant, Carolyn (1980) *The Death of Nature: Women, Ecology and the Scientific Revolution.* San Francisco: Harper & Row Publishers.

Merton, Robert K. (1973) *The Sociology of Science: Theoretical and Empirical Investigations.* Ed. N. W. Storer. Chicago: University of Chicago Press.

Merz, Martina (1999) "Multiplex and Unfolding: Computer Simulation in Particle Physics. *Science in Context* 12: 293–316.

Merz, Martina and Karin Knorr Cetina (1997) "Deconstruction in a 'Thinking' Science: Theoretical Physicists at Work." *Social Studies of Science* 27: 73–112.

Miettinen, Reijo (1998) "Object Construction and Networks in Research Work: The Case of Research on Cullulose-Degrading Enzymes." *Social Studies of Science* 28: 423–64.

Miller, Clark (2004) "Interrogating the Civic Epistemology of American Democracy: Stability and Instability in the 2000 US Presidential Election". *Social Studies of Science* 34: 501–30.

Mirowski, Philip (1989) *More Heat than Light: Economics as Social Physics, Physics as Nature's Economics.* Cambridge: Cambridge University Press.

Mirowski, Philip and Esther-Mirjam Sent (2007) "The Commercialization of Science and the Response of STS." In E. J. Hackett, O. Amsterdamska, M. Lynch and J. Wajcman, eds, *The Handbook of Science and Technology Studies*, 3rd edn. Cambridge, MA: MIT Press, 635–89.

Mirowski, Philip and Robert Van Horn (2005) "The Contract Research Organization and the Commercialization of Scientific Research." *Social Studies of Science* 35: 503–34.

Misa, Thomas J. (1992) "Controversy and Closure in Technological Change: Constructing 'Steel'." In W. Bijker and J. Law, eds, *Shaping Technology/Building Society: Studies in Sociotechnical Change.* Cambridge, MA: MIT Press, 109–39.

Misak, C. J. (1995) *Verificationism: Its History and Prospects* (1995 edn). New York: Routledge.

Mitcham, Carl (1994) *Thinking through Technology: The Path between Engineering and Philosophy.* Chicago: Chicago University Press.

Mitroff, Ian I. (1974) "Norms and Counter Norms in a Select Group of the Apollo Moon Scientists: A Case Study of the Ambivalence of Scientists." *American Sociological Review* 39: 579–95.

Mohr, Alison (2002) "Of Being Seen to do the Right Thing: Provisional Findings from the First Australian Consensus Conference on Gene Technology in the Food Chain." *Science and Public Policy* 29: 2–12.

Mol, Annemarie (2002) *The Body Multiple: Ontology in Medical Practice.* Durham, NC: Duke University Press.

Moore, Kelly (1996) "Organizing Integrity: American Science and the Creation of Public Interest Organizations, 1955–1975." *American Journal of Sociology* 101: 1592–1627.

Moore, Lisa Jean (1997) " 'It's Like You Use Pots and Pans to Cook. It's the Tool': The Technologies of Safer Sex." *Science, Technology & Human Values* 22: 434–71.

Moravcsik, Michael J. and J. M. Ziman (1975) "Paradisia and Dominatia: Science and the Developing World." *Foreign Affairs* 53: 699–724.

Mukerji, Chandra (1989) *A Fragile Power: Scientists and the State.* Princeton: Princeton University Press.

Mukerji, Chandra (2006) "Tacit Knowledge and Classical Technique in Seventeenth-Century France: Hydraulic Cement as a Living Practice among Masons and Military Engineers." *Technology and Culture* 47: 713–33.

Mulkay, Michael (1969) "Some Aspects of Cultural Growth in the Sciences." *Social Research* 36: 22–52.

Mulkay, Michael (1980) "Interpretation and the Use of Rules: The Case of Norms of Science." In T. F. Gieryn, ed., *Science and Social Structure: A Festschrift for Robert K. Merton* (Transactions of the New York Academy of Sciences, Series II, Vol. 39). New York: New York Academy of Sciences, 111–25.

Mulkay, Michael (1985) *The Word and the World: Explorations in the Form of Sociological Analysis.* London: George Allen & Unwin.

Mulkay, Michael (1989) "Looking Backward." *Science, Technology & Human Values* 14: 441–59.

Mulkay, Michael (1994) "The Triumph of the Pre-Embryo: Interpretations of the Human Embryo in Parliamentary Debate over Embryo Research." *Social Studies of Science* 24: 611–39.

Mumford, Lewis (1934) *Technics and Civilization* (1934 edn). New York: Harcourt Brace.

Mumford, Lewis (1967) *The Myth of the Machine* (Vol. 1). New York: Harcourt Brace Jovanovich.

Murphy, Michelle (2006) *Sick Building Syndrome and the Problem of Uncertainty.* Durham, NC: Duke University Press.

Myers, Greg (1990) *Writing Biology: Texts in the Social Construction of Scientific Knowledge.* Madison: University of Wisconsin Press.

Myers, Greg (1995) "From Discovery to Invention: The Writing and Rewriting of Two Patents." *Social Studies of Science* 25: 57–106.

Myers, Natasha (2008) "Molecular Embodiments and the Body-work of Modeling in Protein Crystallography." *Social Studies of Science* 38: 163–99.

Nagel, Thomas (1989) *The View from Nowhere.* Oxford: Oxford University Press.

Nandy, Ashis (1988) *Science, Hegemony and Violence: A Requiem for Modernity.* Tokyo: The United Nations University.

National Academy of Sciences (1995) *On Being a Scientist*, 2nd edn. Washington, DC: National Academy Press.

National Science Board (2008) *Science and Engineering Indicators 2008.* Arlington, VA: National Science Foundation (Vol. 1, NSB 08-01; Vol. 2, NSB 08-01A).

Nelkin, Dorothy (1995) *Selling Science: How the Press Covers Science and Technology*, 2nd edn. New York: W. H. Freeman.

Nelkin, Dorothy and M. Susan Lindee (1995) *The DNA Mystique: The Gene as a Cultural Icon.* New York: W. H. Freeman.

Nersession, Nancy J. (1988) "Reasoning from Imagery and Analogy in Scientific Concept Formation." In A. Fine and J. Leplin, eds, *PSA 1988: Proceedings of the 1988 Biennial Meeting of the Philosophy of Science Association* (Vol. 1). East Lansing, Michigan: Philosophy of Science Association, 41–7.

Netz, Reviel (1999) *The Shaping of Deduction in Greek Mathematics: A Study in Cognitive History.* Cambridge: Cambridge University Press.

Nishizawa, Mariko (2005) "Citizen Deliberations on Science and Technology and their Social Environments: Case Study on the Japanese Consensus Conference on GM crops." *Science and Public Policy* 32: 479–89.

Noble, David F. (1984) *Forces of Production: A Social History of Industrial Automation*. New York: Oxford University Press.

Noble, David F. (1992) *A World Without Women: The Christian Clerical Culture of Western Science*. New York: Alfred A. Knopf.

Nowotny, Helga, Peter Scott, and Michael Gibbons (2001) *Re-Thinking Science: Knowledge and the Public in an Age of Uncertainty*. London: Polity Press.

Nowotny, Helga (2005) "The Changing Nature of Public Science." In H. Nowotny, D. Petre, E. Schmidt-Assmann, H. Schulze-Fielitz, and H.-H. Trute, eds, *The Public Nature of Science under Assault: Politics, Markets, Science and the Law*. Berlin: Springer, 1–27.

O'Connell, Joseph (1993) "Metrology: The Creation of Universality by the Circulation of Particulars." *Social Studies of Science* 23: 129–74.

Orenstein, Peggy (1994) *School Girls*. New York: Doubleday.

Oreskes, Naomi and Erik M. Conway (2008) "Challenging Knowledge: How Climate Science Became a Victim of the Cold War." In R. N. Proctor and L. Schiebinger, eds, *Agnotology: The Making and Unmaking of Ignorance*. Stanford: Stanford University Press, 55–89.

Oudshoorn, Nelly (1994) *Beyond the Natural Body: An Archaeology of Sex Hormones*. London: Routledge.

Oudshoorn, Nelly (2003) *The Male Pill: A Biography of a Technology in the Making*. Durham, NC: Duke University Press.

Oudshoorn, Nelly and Trevor Pinch, eds (2003) *How Users Matter: The Co-construction of Users and Technology*. Cambridge, MA: MIT Press.

Packer, Kathryn and Andrew Webster (1996) "Patenting Culture in Science: Reinventing the Scientific Wheel of Credibility." *Science, Technology & Human Values* 21: 427–53.

Papineau, David, ed. (1996) *The Philosophy of Science*. New York: Oxford University Press.

Pauly, Philip J. (1996) "The Beauty and Menace of the Japanese Cherry Trees: Conflicting Visions of American Ecological Independence." *Isis* 87: 51–73.

Perelman, Chaim and Lucie Olbrechts-Tyteca (1969) *The New Rhetoric: A Treatise on Argumentation*. Tr. J. Wilkinson and P. Weaver. Notre Dame: University of Notre Dame Press.

Petryna, Adriana (2007) "Clinical Trials Offshored: On Private Sector Science and Public Health." *BioSocieties* 2: 21–40.

Philip, Kavita (1995) "Imperial Science Rescues a Tree: Global Botanic Networks, Local Knowledge and the Transcontinental Transplantation of Cinchona." *Environment and History* 1: 173–200.

Phillips, David M. (1991) "Importance of the Lay Press in the Transmission of Medical Knowledge to the Scientific Community." *New England Journal of Medicine* (11 October): 1180–83.

Picart, Caroline Joan S. (1994) "Scientific Controversy as Farce: The Benveniste–Maddox Counter Trials." *Social Studies of Science* 24: 7–38.

Pickering, Andrew (1984) *Constructing Quarks: A Sociological History of Particle Physics*. Chicago: University of Chicago Press.

Pickering, Andrew (1992) "From Science as Knowledge to Science as Practice." In A. Pickering, ed., *Science as Practice and Culture*. Chicago: University of Chicago Press, 1–26.

Pickering, Andrew, ed. (1992) *Science as Practice and Culture*. Chicago: University of Chicago Press.

Pickering, Andrew (1995) *The Mangle of Practice: Time, Agency, and Science*. Chicago: University of Chicago Press.

Pielke, Roger A., Jr. (2007) *The Honest Broker: Making Sense of Science in Policy and Politics*. Cambridge: Cambridge University Press.

Pinch, Trevor (1985) "Towards an Analysis of Scientific Observation: The Externality and Evidential Significance of Observational Reports in Physics." *Social Studies of Science* 15: 3–36.

Pinch, Trevor (1993a) " 'Testing – One, Two, Three . . . Testing!' Toward a Sociology of Testing." *Science, Technology & Human Values* 18: 25–41.

Pinch, Trevor (1993b) "Turn, Turn, and Turn Again: The Woolgar Formula." *Science, Technology & Human Values* 18: 511–522.

Pinch, Trevor J. and Wiebe E. Bijker (1987) "The Social Construction of Facts and Artifacts: Or How the Sociology of Science and the Sociology of Technology Might Benefit Each Other." In W. E. Bijker, T. P. Hughes, and T. Pinch, eds, *The Social Construction of Technological Systems: New Directions in the Sociology and History of Technology*. Cambridge, MA: The MIT Press, 17–50.

Pinch, Trevor and Karin Bijsterveld (2004) "Sound Studies: New Technologies and Music." *Social Studies of Science* 34: 635–48.

Polanyi, Michael (1958) *Personal Knowledge: Towards a Post-Critical Philosophy*. Chicago: University of Chicago Press.

Polanyi, Michael (1962) "The Republic of Science: Its Political and Economic Theory." *Minerva* 1: 54–73.

Popper, Karl (1963) *Conjectures and Refutations: The Growth of Scientific Knowledge*. London: Routledge & Kegan Paul.

Porter, Theodore M. (1992a) "Objectivity as Standardization: The Rhetoric of Impersonality in Measurement, Statistics, and Cost–Benefit Analysis." *Annals of Scholarship* 9: 19–59.

Porter, Theodore M. (1992b) "Quantification and the Accounting Ideal in Science." *Social Studies of Science* 22: 633–52.

Porter, Theodore M. (1995) *Trust in Numbers: The Pursuit of Objectivity in Science and Public Life*. Princeton: Princeton University Press.

Prasad, Amit (2007) "The (Amorphous) Anatomy of an Invention: The Case of Magnetic Resonance Imaging (MRI)." *Social Studies of Science* 37: 533–60.

Prelli, Lawrence J. (1989) *A Rhetoric of Science: Inventing Scientific Discourse*. Columbia, SC: University of South Carolina Press.

Proctor, Robert N. (1991) *Value-Free Science? Purity and Power in Modern Knowledge*. Cambridge, MA: Harvard University Press.

Proctor, Robert N. (2008) "Agnotology: A Missing Term to Describe the Cultural Production of Ignorance (and Its Study)." In R. N. Proctor and L. Schiebinger, eds, *Agnotology: The Making and Unmaking of Ignorance*. Stanford: Stanford University Press, 1–33.

Proctor, Robert N. and Londa Schiebinger, eds (2008) *Agnotology: The Making and Unmaking of Ignorance*. Stanford: Stanford University Press.

Putnam, Hilary (1981) "The 'Corroboration' of Theories." In I. Hacking, ed., *Scientific Revolutions*. Oxford: Oxford University Press, 60–79.

Pyenson, Lewis (1985) *Cultural Imperialism and Exact Sciences: German Expansion Overseas*. New York: Peter Lang.

Rabinow, Paul and Nikolas Rose (2006) "Biopower Today." *Biosocieties* 1: 195–217.

Radder, Hans (1988) *The Material Realization of Science: A Philosophical View on the Experimental Natural Sciences, Developed in Discussion with Habermas*. Assen/Maastricht: Van Gorcum.

Radder, Hans (1993) "Science, Realization and Reality: The Fundamental Issues." *Studies in History and Philosophy of Science* 24: 327–49.

Rader, Karen (1998) " 'The Mouse People': Murine Genetics Work at the Bussey Institution, 1909–1936." *Journal of the History of Biology* 31: 327–54.

Rajan, Kaushik Sunder (2006) *Biocapital: The Constitution of Postgenomic Life*. Durham, NC: Duke University Press.

Rappert, Brian (2001) "The Distribution and Resolutions of the Ambiguities of Technology, or Why Bobby Can't Spray." *Social Studies of Science* 31: 557–91.

Ravetz, Jerome R. (1971) *Scientific Knowledge and its Social Problems*. Oxford: Clarendon.

Reardon, Jenny (2001) "The Human Genome Diversity Project: A Case Study in Coproduction." *Social Studies of Science* 31: 357–88.

Restivo, Sal (1990) "The Social Roots of Pure Mathematics." In S. E. Cozzens and T. F. Gieryn, eds, *Theories of Science in Society*. Bloomington: Indiana University Press, 120–43.

Rheinberger, Hans-Jorg (1997) *Toward a History of Epistemic Things: Synthesizing Proteins in the Test Tube*. Stanford, CA: Stanford University Press.

Richards, Evelleen (1991) *Vitamin C and Cancer: Medicine or Politics?* London: Macmillan.

Richards, Evelleen (1996) "(Un)Boxing the Monster." *Social Studies of Science* 26: 323–56.

Richards, Evelleen and John Schuster (1989) "The Feminine Method as Myth and Accounting Resource: A Challenge to Gender Studies and Social Studies of Science." *Social Studies of Science* 19: 697–720.

Richardson, Alan W. (1998) *Carnap's Construction of the World: The Aufbau and the Emergence of Logical Empiricism*. Cambridge: Cambridge University Press.

Richmond, Marsha L. (1997) " 'A Lab of One's Own': The Balfour Biological Laboratory for Women at Cambridge University, 1884–1914." *Isis* 88: 422–55.

Robertson, Frances (2005) "The Aesthetics of Authenticity: Printed Banknotes as Industrial Currency." *Technology and Culture* 46: 31–50.

Rose, Hilary (1986) "Beyond Masculinist Realities: A Feminist Epistemology for the Sciences." In R. Bleier, ed., *Feminist Approaches to Science*. New York: Pergamon Press.

Rosen, Paul (1993) "Social Construction of Mountain Bikes: Technology and Post-modernity in the Cycle Industry." *Social Studies of Science* 23: 479–514.

Rosenberg, Nathan (1994) *Exploring the Black Box: Technology, Economics, and History*. Cambridge: Cambridge University Press.

Ross, Joseph S., Kevin P. Hill, David S. Egilman, and Harlan M. Krumholz (2008) "Guest Authorship and Ghostwriting in Publications Related to Rofecoxib: A Case Study of Industry Documents from Rofecoxib Litigation." *Journal of the American Medical Association* 299 (15): 1800–12.

Rossiter, Margaret (1982) *Women Scientists in America: Struggles and Strategies to 1940*. Baltimore: Johns Hopkins University Press.

Rossiter, Margaret (1995) *Women Scientists in America: Before Affirmative Action*. Baltimore: Johns Hopkins University Press.

Roth, Wolff-Michael and G. Michael Bowen (1999) "Digitizing Lizards: The Topology of 'Vision' in Ecological Fieldwork." *Social Studies of Science* 29: 719–64.

Rowe, Gene and Lynn J. Frewer (2005) "A Typology of Public Engagement Mechanisms." *Science, Technology & Human Values* 30: 251–90.

Rowe, Gene, Roy Marsh, and Lynn J. Frewer (2004) "Evaluation of a Deliberative Conference." *Science, Technology & Human Values* 29: 88–121.

Rudwick, Martin J. S. (1974) "Poulett Scrope on the Volcanoes of Auvergne: Lyellian Time and Political Economy." *The British Journal for the History of Science* 7: 205–42.

Salonius, Annalisa (2009) "Social Organization of Work in Biomedical Research Labs in Leading Universities in Canada: Socio-historical Dynamics and the Influence of Research Funding." *Social Studies of Science* 39 (forthcoming).

Schiebinger, Londa (1993) *Nature's Body: Gender in the Making of Modern Science*. Boston: Beacon Press.

Schiebinger, Londa (1999) *Has Feminism Changed Science?* Cambridge, MA: Harvard University Press.

Sclove, Richard (1995) *Democracy and Technology*. New York: Guilford Press.

Sclove, Richard (2000) "Town Meetings on Technology." In Daniel Lee Kleinman, ed., *Science, Technology and Democracy*. Albany, NY: SUNY Press, 33–48.

Scott, Pam, Evelleen Richards, and Brian Martin (1990) "Captives of Controversy: The Myth of the Neutral Social Researcher in Contemporary Scientific Controversies." *Science, Technology & Human Values* 15: 474–94.

Secord, James A. (1981) "Nature's Fancy: Charles Darwin and the Breeding of Pigeons." *Isis* 72: 163–86.

Selzer, Jack, ed. (1993) *Understanding Scientific Prose*. Madison: University of Wisconsin Press.

Sen, Amartya (2000) *Development as Freedom*. Oxford: Oxford University Press.

Sengoopta, Chandak (1998) "Glandular Politics: Experimental Biology, Clinical Medicine, and Homosexual Emancipation in Fin-de-Siecle Central Europe." *Isis* 89: 445–73.

Shackley, Simon and Brian Wynne (1995) "Global Climate Change: The Mutual Construction of an Emergent Science-Policy Domain." *Science and Public Policy* 22: 218–30.

Shapin, Steven (1975) "Phrenological Knowledge and the Social Structure of Early Nineteenth-Century Edinburgh." *Annals of Science* 32: 219–43.

Shapin, Steven (1981) "Of Gods and Kings: Natural Philosophy and Politics in the Leibniz–Clarke Disputes." *Isis* 72: 187–215.

Shapin, Steven (1988) "The House of Experiment in Seventeenth-Century England." *Isis* 79: 373–404.

Shapin, Steven (1994) *A Social History of Truth: Civility and Science in Seventeenth-Century England*. Chicago: University of Chicago Press.

Shapin, Steven and Simon Schaffer (1985) *Leviathan and the Air-Pump: Hobbes, Boyle, and the Experimental Life*. Princeton: Princeton University Press.

Sharif, Naubuhar (2006) "Emergence and Development of the National Innovation Systems Concept." *Research Policy* 35: 745–66.

Shenhav, Yehouda and David Kamens (1991) "The Cost of Institutional Isomorphism: Science in Less Developed Countries." *Social Studies of Science* 21: 527–45.

Shepherd, Chris and Martin R. Gibbs (2006) " 'Stretching the Friendship': On the Politics of Replicating a Dairy in East Timor." *Science, Technology & Human Values* 31: 668–701.

Shiva, Vandana (1997) *Biopiracy: The Plunder of Nature and Knowledge*. Boston: South End Press.

Shrum, Wesley (2000) "Science and Story in Development: The Emergence of Non-Governmental Organizations in Agricultural Research." *Social Studies of Science* 30: 95–124.

Shrum, Wesley and Yehouda Shenhav (1995) "Science and Technology in Less Developed Countries." In S. Jasanoff, G. E. Markle, J. C. Petersen, and T. Pinch, eds, *Handbook of Science and Technology Studies*. London: Sage, 627–51.

Simon, Bart (1999) "Undead Science: Making Sense of Cold Fusion After the (Arti)fact." *Social Studies of Science* 29: 61–86.

Singleton, Vicky and Mike Michael (1993) "Actor-networks and Ambivalence: General Practitioners in the Cervical Screening Programme." *Social Studies of Science* 23: 227–64.

Sismondo, Sergio (1995) "The Scientific Domains of Feminist Standpoints." *Perspectives on Science* 3: 49–65.

Sismondo, Sergio (1996) *Science without Myth: On Constructions, Reality, and Social Knowledge*. Albany, NY: SUNY Press.

Sismondo, Sergio (1997) "Modelling Strategies: Creating Autonomy for Biology's Theory of Games." *History and Philosophy of the Life Sciences* 19: 147–61.

Sismondo, Sergio (2000) "Island Biogeography and the Multiple Domains of Models." *Biology and Philosophy* 15: 239–258.

Sismondo, Sergio (2009) "Ghosts in the Machine: Publication Planning in the Medical Sciences." *Social Studies of Science* 39: 171–98.

Slaughter, Sheila and Larry L. Leslie (1997) *Academic Capitalism: Politics, Policies, and the Entrepreneurial University*. Baltimore: Johns Hopkins University Press.

Slaughter, Sheila and Gary Rhoades (2004) *Academic Capitalism and the New Economy: Markets, State, and Higher Education.* Baltimore: Johns Hopkins University Press.

Smith, Dorothy (1987) *The Everyday World as Problematic: A Feminist Sociology.* Boston: Northeastern University Press.

Smithson, Michael J. (2008) "Social Theories of Ignorance." In R. N. Proctor and L. Schiebinger, eds, *Agnotology: The Making and Unmaking of Ignorance.* Stanford: Stanford University Press, 209–29.

Sørensen, Knut H. (1992) "Towards a Feminized Technology? Gendered Values in the Construction of Technology." *Social Studies of Science* 22: 5–31.

Sørensen, Knut H. and Anne-Jorunn Berg (1987) "Genderization of Technology Among Norwegian Engineering Students." *Acta Sociologica* 30: 151–71.

Soulé, Michael and Gary Lease, eds (1995) *Reinventing Nature? Responses to Postmodern Deconstruction.* Washington, DC: Island Press.

Spiegel-Rösing, Ina and Derek de Solla Price, eds (1977) *Science, Technology and Society: A Cross-Disciplinary Perspective.* London: Sage.

Star, Susan Leigh (1983) "Simplification in Scientific Work: An Example from Neuro-science Research." *Social Studies of Science* 13: 205–28.

Star, Susan Leigh (1989) "Layered Space, Formal Representations, and Long-Distance Control: The Politics of Information." *Fundamenta Scientiae* 10: 125–54.

Star, Susan Leigh (1991) "Power, Technologies and the Phenomenology of Conventions: On Being Allergic to Onions." In J. Law, ed., *A Sociology of Monsters: Essays on Power, Technology and Domination.* London: Routledge, 26–56.

Star, Susan Leigh (1992) "Craft vs. Commodity, Mess vs. Transcendence: How the Right Tool Became the Wrong One in the Case of Taxidermy and Natural History." In A. Clarke and J. Fujimura, eds, *The Right Tools for the Job: At Work in Twentieth-Century Life Sciences.* Princeton: Princeton University Press, 257–86.

Star, Susan Leigh and James R. Griesemer (1989) "Institutional Ecology, 'Translations' and Boundary Objects: Amateurs and Professionals in Berkeley's Museum of Vertebrate Zoology, 1907–39." *Social Studies of Science* 19: 387–420.

Stirling, Andy (2008) " 'Opening Up' and 'Closing Down': Power, Participation, and Pluralism in the Social Appraisal of Technology." *Science Technology & Human Values* 33: 262–94.

Stocking, S. Holly and Lisa W. Holstein (2008) "Manufacturing Doubt: Journalists' Roles and the Construction of Ignorance in a Scientific Controversy." *Public Understanding of Science* 18: 23–42.

Stokes, Donald E. (1997) *Pasteur's Quadrant: Basic Science and Technological Innovation.* Washington, DC: Brookings Institution Press.

Sturgis, Patrick and Nick Allum (2004) "Science in Society: Re-Evaluating the Deficit Model of Public Attitudes." *Public Understanding of Science* 13: 55–74.

Sundberg, Mikaela (2007) "Parameterizations as Boundary Objects on the Climate Arena." *Social Studies of Science* 37: 473–88.

Takacs, David (1996) *The Idea of Biodiversity: Philosophies of Paradise*. Baltimore: Johns Hopkins University Press.

Taylor, Peter (1995) "Co-Construction and Process: A Response to Sismondo's Classification of Constructivisms." *Social Studies of Science* 25: 348–59.

Thompson, Charis (2005) *Making Parents: The Ontological Choreography of Reproductive Technologies*. Cambridge, MA: MIT Press.

Tilghman, Shirley (1998) "Science vs. The Female Scientist." In J. Kourany, ed., *Scientific Knowledge: Basic Issues in the Philosophy of Science*. Belmont, CA: Wadsworth Publishing Company, 40–42.

Toulmin, Stephen (1958) *The Uses of Argument*. Cambridge: Cambridge University Press.

Toulmin, Stephen (1990) *Cosmopolis: The Hidden Agenda of Modernity*. Chicago: University of Chicago Press.

Traweek, Sharon (1988) *Beamtimes and Lifetimes: The World of High Energy Physicists*. Cambridge, MA: Harvard University Press.

Tribby, Jay (1994) "Club Medici: Natural Experiment and the Imagineering of 'Tuscany'." *Configurations* 2: 215–35.

Tufte, Edward (1997) *Visual Explanations: Images and Quantities, Evidence and Narrative*. Cheshire, CT: Graphics Press.

Turkle, Sherry (1984) *The Second Self: Computers and the Human Spirit*. New York: Simon & Schuster.

Turkle, Sherry and Seymour Papert (1990) "Epistemological Pluralism: Styles and Voices within the Computer Culture." *Signs* 16: 128–57.

Turnbull, David (1995) "Rendering Turbulence Orderly." *Social Studies of Science* 25: 9–34.

Turner, S. P. and Chubin, D. E. (1976) "Another Appraisal of Ortega, the Coles and Science Policy: The Ecclesiastes Hypothesis." *Social Science Information* 15: 657–62.

Turner, Stephen (2003) *Liberal Democracy 3.0: Civil Society in an Age of Experts*. London: Sage.

Urry, John (2004) "The 'System' of Automobility." *Theory, Culture & Society* 21 (4/5): 25–39.

Van den Daele, Wolfgang (1977) "The Social Construction of Science: Institutionalisation and Definition of Positive Science in the Latter Half of the Seventeenth Century." In E. Mendelsohn, P. Weingart, and R. Whitley, eds, *The Social Production of Scientific Knowledge*. Dordrecht: D. Reidel, 27–54.

van Fraassen, Bas C. (1980) *The Scientific Image*. Oxford: Clarendon Press.

Varma, Roli (2000) "Changing Research Cultures in U.S. Industry." *Science, Technology & Human Values* 25: 395–416.

Vaughan, Diane (1996) *The Challenger Launch Decision: Risky Technology, Culture, and Deviance at NASA*. Chicago: University of Chicago Press.

Vincenti, Walter G. (1990) *What Engineers Know and How They Know It: Analytical Studies from Aeronautical History*. Baltimore: Johns Hopkins University Press.

Vincenti, Walter G. (1995) "The Technical Shaping of Technology: Real-World Constraints and Technical Logic in Edison's Electrical Lighting System." *Social Studies of Science* 25: 553–74.

von Hippel, Eric (2005) *Democratizing Innovation.* Cambridge, MA: MIT Press.

Wajcman, Judy (1991) *Feminism Confronts Technology.* University Park, PA: The Pennsylvania State University Press.

Waldby, Catherine and Robert Mitchell (2006) *Tissue Economies: Blood, Organs, and Cell Lines in Late Capitalism.* Durham, NC: Duke University Press.

Warwick, Andrew (2003) *Masters of Theory: Cambridge and the Rise of Mathematical Physics.* Chicago: University of Chicago Press.

Weber, Rachel N. (1997) "Manufacturing Gender in Commercial and Military Cockpit Design." *Science, Technology & Human Values* 22: 235–53.

Wenneras, Christine and Agnes Wold (2001) "Nepotism and Sexism in Peer-Review." In M. Lederman and I. Bartsch, eds, *The Gender and Science Reader.* London: Routledge.

Whitehead, Alfred North (1992) *An Introduction to Mathematics* (1st edn 1911). Oxford: Oxford University Press.

Winner, Langdon (1986a) "Do Artifacts Have Politics?" in Langdon Winner, *The Whale and the Reactor: A Search for Limits in an Age of High Technology.* Chicago: University of Chicago Press, 19–39.

Winner, Langdon (1986b) *The Whale and the Reactor: A Search for Limits in an Age of High Technology.* Chicago: University of Chicago Press.

Wittgenstein, Ludwig (1958) *Philosophical Investigations,* 2nd edn. Tr. G. E. M. Anscombe. Oxford: Blackwell.

Wood, S. (1982) *The Degradation of Work? Skill, Deskilling and the Labour Process.* London: Hutchinson.

Woolgar, Steve (1981) "Interests and Explanation in the Social Study of Science." *Social Studies of Science* 11: 365–94.

Woolgar, Steve (1988) *Science: The Very Idea.* Chichester, Sussex: Ellis Horwood.

Woolgar, Steve (1992) "Some Remarks about Positionism: A Reply to Collins and Yearley." In A. Pickering, ed., *Science as Practice and Culture.* Chicago: University of Chicago Press, 327–42.

Woolgar, Steve (1993) "What's at Stake in the Sociology of Technology? A Reply to Pinch and to Winner." *Science, Technology & Human Values* 18: 523–29.

Wray, K. Brad (2003) "Is Science Really a Young Man's Game?" *Social Studies of Science* 33: 137–49.

Wright, Nick and Brigitte Nerlich (2006) "Use of the Deficit Model in a Shared Culture of Argumentation: The Case of Foot and Mouth Science." *Public Understanding of Science* 15: 331–42.

Wyatt, Sally (2004) "Danger! Metaphors at Work in Economics, Geophysiology, and the Internet." *Science Technology & Human Values* 29: 242–61.

Wyatt, Sally (2007) "Technological Determinism is Dead. Long Live Technological Determinism." In E. J. Hackett, O. Amsterdamska, M. Lynch and J. Wajcman, eds, *The Handbook of Science and Technology Studies*, 3rd edn. Cambridge, MA: MIT Press, 165–80.

Wynne, Brian (1992) "Public Understanding of Science Research: New Horizons or Hall of Mirrors?" *Public Understanding of Science* 1: 37–43.

Wynne, Brian (1996) "May the Sheep Safely Graze? A Reflexive View of the Expert-Lay Knowledge Divide." In S. Lash, B. Szerszynski, and B. Wynne, eds, *Risk, Environment & Modernity*. London: Sage, 44–83.

Wynne, Brian (2003) "Seasick on the Third Wave? Subverting the Hegemony of Propositionalism." *Social Studies of Science* 33: 401–18.

Yearley, Steven (1999) "Computer Models and the Public's Understanding of Science: A Case-Study Analysis." *Social Studies of Science* 29: 845–66.

Ylikoski, Petri (2001) *Understanding Interests and Causal Explanation*. Ph.D. Thesis, University of Helsinki.

Zabusky, Stacia E. and Stephen R. Barley (1997) " 'You Can't be a Stone if You're Cement': Reevaluating the Emic Identities of Scientists in Organizations." *Organizational Behavior* 19: 361–404.

Zenzen, Michael and Sal Restivo (1982) "The Mysterious Morphology of Immiscible Liquids: A Study of Scientific Practice." *Social Science Information* 21: 447–73.

Ziman, John (1984) *An Introduction to Science Studies: The Philosophical and Social Aspects of Science and Technology*. Cambridge: Cambridge University Press.

Ziman, John (1994) *Prometheus Bound: Science in a Dynamic Steady State*. Cambridge: Cambridge University Press.

Zuckerman, Harriet (1977) "Deviant Behavior and Social Control in Science." In E. Sagarin, ed., *Deviance and Social Change*. London: Sage, 87–138.

Zuckerman, Harriet (1984) "Norms and Deviant Behavior in Science." *Science, Technology & Human Values* 9: 7–13.

Index

CPSIA information can be obtained
at www.ICGtesting.com
Printed in the USA
LVHW080719171218
600707LV00012BA/317/P